Communications in Computer and Information Science 603

Commenced Publication in 2007
Founding and Former Series Editors:
Alfredo Cuzzocrea, Dominik Ślęzak, and Xiaokang Yang

Editorial Board

More information about this series at http://www.springer.com/series/7899

Syng Yup Ohn · Sung Do Chi (Eds.)

Model Design and Simulation Analysis

15th International Conference, AsiaSim 2015
Jeju, Korea, November 4–7, 2015
Revised Selected Papers

Springer

Editors
Syng Yup Ohn
Korea Aerospace University
Goyang-si
Korea (Republic of)

Sung Do Chi
Korea Aerospace University
Goyang-si
Korea (Republic of)

ISSN 1865-0929 ISSN 1865-0937 (electronic)
Communications in Computer and Information Science
ISBN 978-981-10-2157-2 ISBN 978-981-10-2158-9 (eBook)
DOI 10.1007/978-981-10-2158-9

Library of Congress Control Number: 2016946015

Preface

AsiaSim (Asia Simulation Conference) is an annual conference organized by the AsiaSim Federation member society in cooperation with the Asia Simulation Federation. Since 2001, AsiaSim has provided a forum for scientists, academics, and professionals from around Asia to present their latest and exciting research findings in various fields of modeling, simulation, and their applications. This strict screening process results in the presentation of the high-quality papers in the fields of modeling and simulation methodology, virtual reality, aerospace, e-business, manufacturing, medical, military, networks, transportation, and traffic simulation and general engineering applications.

AsiaSim 2015 received 126 submissions. After a thorough reviewing process, 71 papers were selected for oral presentations and 32 papers were selected for posters. Among them, only 11 papers were finally accepted for CCIS as full papers (acceptance ratio of 8.73 %). This volume contains the revised full version of 11 English papers presented at AsiaSim 2015.

The high-quality program would not have been possible without the authors who chose AsiaSim 2015 as a venue for their publications. We are also very grateful to the Program Committee members and Organizing Committee members, who put a tremendous amount of effort into soliciting and selecting research papers with a balance of high quality, new ideas and new novel applications. We also thank the external reviewers for their time, effort, and timely response.

We wish to express our special thanks to General Chair Prof. Yun Bae Kim (Sungkyunkwan University, Korea) for his advice on all aspects of the conference. Thank you also to the attendees for joining us at AsiaSim 2015.

June 2016   Seon Yep Ohn
                                                                             Sung Do Chi

Organization

Honorary Chairs

Jong-Hyun Kim	Yonsei University, Korea
Axel Lehmann	Universität der Bundeswehr München, Germany
Bo Hu Li	Beihang University, China
Koji Koyamada	Kyoto University, Japan

General Chair

Yun Bae Kim	Sungkyunkwan University, Korea

General Co-chairs

Zhao Qinping	Beihang University, President of the CASS, China
Satoshi Tanaka	Ritsumeikan University, President of the ASIASIM and the JSST, Japan

Organizing Committee

Seong-Yong Jang (Chair)	Seoul National University of Science and Technology, Korea
Byeong-Yun Chang (Chair)	Ajou University, Korea
Jinsoo Park	Yong In University, Korea
Chul-Jin Park	Hanyang University, Korea

International Program Committee

Jong-Sik Lee (Chair)	Inha University, Korea
Hae Young Lee	Seoul Women's University, Korea
Lin Zhang	Beihang University, China
Xiaofeng Hu	Beihang University, China
Yunjie Wu	Beihang University, China
Ming Yang	Beihang University, China
Osamu Ono	Meiji University, Japan
Kyoko Hasegawa	Ritsumeikan University, Japan
Gary Tan	National University of Singapore, Singapore
Rubiyah Yusof	Universiti Teknologi Malaysia, Malaysia
Yong Meng Teo	National University of Singapore, Singapore
Yahaya Md Sam	Universiti Teknologi Malaysia, Malaysia
Xiao Song	Beihang University, China

Publication Committee

Dong-Won Seo (Proceedings Chair)	Kyunghee University, Korea
Ja-Hee Kim	Seoul National University of Science and Technology, Korea
Sung-Do Chi (Special Issue Chair)	Korea Aerospace University, Korea
Yun-Ho Seo (Special Issue Chair)	Korea University, Korea
Jang-Se Lee	Korea Maritime and Ocean University, Korea
Dong-Hyun Choi	Korea Aerospace University, Korea
Sol Ha	Seoul National University, Korea

Industrial Committee

Dug-Hee Moon (Chair)	Changwon National University, Korea
Byung Hee Kim	VMS SOLUTIONS Co., Ltd., Korea
Seung Whan Kim	Siemens Industry Software Ltd, Korea
Young Suk Park	ATWORTH CO., LTD., Korea
Ku-kil Chang	Dassault Systems Korea, Korea
Young Gyo Chung	SimTech Systems, Inc., Korea
Seong-Hoon Choi	Sangmyung University, Korea

Award Committee

Hyung-Jong Kim (Chair)	Seoul Women's University, Korea

International Advisory Committee

Yun Bae Kim (Chair)	Sungkyunkwan University, Korea
Gyu Min Lee	Pusan National University, Korea
Sam-joon Park	Agency of Defense Development, Korea

Contents

X   Contents

Model and Design

A Regularized Finite Volume Numerical Method for the Extended Porous Medium Equation Relevant to Moisture Dynamics with Evaporation in Non-woven Fibrous Sheets

Hidekazu Yoshioka[1]([⊠]) and Dimetre Triadis[2]

[1] Faculty of Life and Environmental Science,
Shimane University, Matsue, Japan
yoshih@life.shimane-u.ac.jp
[2] Institute of Mathematics for Industry, Kyushu University - Australia Branch,
La Trobe University, Melbourne, Australia
D.Triadis@latrobe.edu.au

Abstract. The extended porous medium equation (PME) is a degenerate nonlinear diffusion equation that effectively describes moisture dynamics with evaporation in non-woven fibrous sheets. We propose a new finite volume numerical model of the extended PME incorporating regularization of nonlinear degenerate terms, and apply it to test cases for verification of accuracy, stability, and versatility. One of the test cases considered is a new exact steady solution of the extended PME. We also examine a differential equation-based adaptive re-meshing technique for resolving sharp transitions of solution profiles that may be optionally incorporated into the procedure above. The computational results demonstrate satisfactory accuracy of the proposed numerical model, with reasonable reproduction of complicated moisture dynamics involving sharp transitions and divorce of supports.

Keywords: Moisture dynamics · Evaporation · Non-woven fibrous sheet · Extended porous medium equation · Dual-finite volume method

1 Introduction

Comprehending moisture dynamics occurring in porous media, such as soils, concretes, papers, and fibrous sheets is essential for a wide variety of real systems, and is thus currently an important research topic in hydro-environmental and water resources engineering [1]. Recently, non-woven fibrous sheets have gained attention as cheap and easily obtained materials with a variety of industrial applications, such as agricultural plantation sheets and sanitary products. Assessing moisture dynamics in non-woven fibrous sheets is a crucial step towards creating fit-for-purpose industrial products. As a very recent example in Japan, non-woven fibrous sheets have been used to vertically transport water in a small-scale rooftop vegetation system. This is thought to be a cheap, space-saving, and environmentally friendly plantation system whose source water solely consists of collected rainwater stored in a plastic tank [1]. The resulting

© Springer Science+Business Media Singapore 2016
S.Y. Ohn and S.D. Chi (Eds.): AsiaSim 2015, CCIS 603, pp. 3–16, 2016.
DOI: 10.1007/978-981-10-2158-9_1

moisture dynamics are very complex, and appropriate mathematical and/or numerical models are needed for a practical comprehension of the system.

Moisture dynamics in porous media can be effectively described with Porous Medium Equations (PMEs), an important class of degenerate nonlinear diffusion equations based on the mass conservation principle and constitutive laws [2]. Similar differential equations arise in other applied study areas, such as plasma physics [3] and astrophysics [4]. Solving PMEs involves mathematical and computational difficulties due to degeneracy and nonlinearity of the coefficients. Practical numerical methods for solving PMEs should have sufficiently high accuracy, stability, and versatility. However, such methods are still not common and developing a simple numerical method that is sufficient for practical use is an important research topic.

The main purpose of this article is to present a new numerical model of an extended PME governing longitudinally one-dimensional (1D) moisture dynamics occurring in non-woven fibrous sheets based on regularization, finite volume, and a differential equation-based adaptive re-meshing technique. The proposed numerical model is extensively verified through test cases with exactly or partially known solutions.

2 Mathematical Model

2.1 Extended Porous Medium Equation (PME)

This article considers moisture dynamics in homogenous thin non-woven fibrous sheets with a length scale of 10^{-1} m, width scale of 10^{-2} m, and thickness scale of 10^{-3} m to 10^{-4} m. It is reasonable to assume that transverse variations of the moisture profiles in such fibrous sheets are insignificant compared to longitudinal variations, thus we consider a 1D longitudinal mathematical model. The volumetric water content at each position x of a sheet at the time t is denoted by $\theta = \theta(t, x)$, and is normalized to $0 \leq u = (\theta - \theta_r)(\theta_s - \theta_r)^{-1} \leq 1$ where θ_s and $\theta_r(= 0)$ are the maximum (saturated) and minimum water contents of the sheet. A remarkable difference between modelling moisture dynamics in non-woven fibrous sheets compared with other materials in applications, such as soils and concretes, is that evaporation occurs over the entire 1D or 2D domain in the former case [1] but usually only through the boundary of the domain in the latter case. Considering this characteristic of moisture dynamics in the sheets and adopting dimensionless variables as in the literature [5], the extended PME is proposed in this article as [1, 2, 6]

$$\frac{\partial u}{\partial \tilde{t}} = \frac{\partial}{\partial \tilde{x}} \left((m - p)u^{m-p-1} \frac{\partial u}{\partial \tilde{x}} \right) + \sin \alpha \frac{\partial}{\partial \tilde{x}} \left(u^{m-1} u \right) - \tilde{E}_s u^q \tag{1}$$

with dimensionless variables and parameters

$$\tilde{t} = (m - p)K_s^2 D_s^{-1} \theta_s^{-2} t, \ \tilde{x} = (m - p)K_s D_s^{-1} \theta_s^{-1} x, \ \text{and} \ \tilde{E}_s = (m - p)^{-1} K_s^{-2} D_s \theta_s^2 E_s \tag{2}$$

where $D_s > 0$ is the saturated diffusivity, $K_s > 0$ is the saturated permeability, $E_s \geq 0$ is the evaporation coefficient that depends on both material properties of the sheet and its surrounding environment, m, $p \leq m - 1$, and q are positive nonlinearity parameters, and $-0.5\pi \leq \alpha \leq 0.5\pi$ is the inclination angle of the sheet as defined in Fig. 1.

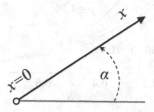

Fig. 1. The 1D domain of the extended PME (1) indicating the inclination α of the fibrous sheet.

In deriving (1), the pressure head $\psi(u)$ and the permeability $K(u)$ have been assumed to be physically parameterized as

$$\psi(u) = p^{-1} K_s^{-1} D_s \theta_s (1 - u^{-p}) \text{ and } K(u) = K_s u^m. \tag{3}$$

Note that the pressure head reduces to the conventional case as $p \to +0$ [2, 6, 7]. The pressure head $\psi(u)$ is continuous and increasing in $0 < u \leq 1$, vanishes at $u = 1$, and diverges as $u \to +0$. The permeability $K(u)$ is continuous and increasing in $0 \leq u \leq 1$ and vanishes at $u = 0$. Figure 2(a) and (b) plot the pressure head $\psi(u)$ and the permeability $K(u)$ for several p and m, respectively where the coefficients $K_s^{-1} D_s \theta_s$ for $\psi(u)$ and K_s for $K(u)$ are set to be 1 for the sake of simplicity. The angle α in (1) may be spatially distributed, but is assumed to be constant in this article. There exist two contrasting cases on arrangement of the sheets, which are the horizontal ($\alpha = 0$) and vertical cases ($\alpha = \pm 0.5\pi$). The advection term of (1) vanishes in the horizontal case ($\sin \alpha = 0$). The standard theory for moisture evaporation from the surface of a soil body involves a near constant evaporation rate for most values of u. It is only as u approaches zero that the evaporation rate rapidly decreases to zero [8]. This behavior has been observed in our preliminary laboratory experiments using thin non-woven fibrous sheets, which are not discussed here but will be addressed in a succeeding paper. In the extended PME, this phenomenon occurs with $q \ll 1$, which has actually been qualitatively validated through preliminary laboratorial experiments. In addition, the order of the parameters m and p have experimentally been estimated as $O = (10^0)$ and $O = (10^{-1})$, respectively, meaning that the relation $p \leq m - 1$ is satisfied for moisture dynamics in non-woven fibrous sheets. Hereafter, "." representing non-dimensional variables are omitted for the sake of brevity.

2.2 Regularized Equation

Degeneration of the advection and diffusion terms of the extended PME (1) would cause severe mathematical and computational issues due to regularity deficits of the coefficients. The nonlinear degeneration of (1) can be mitigated by regularizing it as

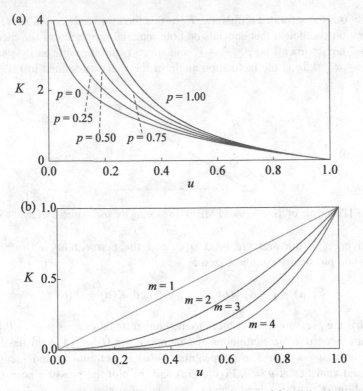

Fig. 2. Plots of (a) the pressure $\psi(u)$ and (b) permeability $K(u)$. (Color figure online)

$$\frac{\partial u}{\partial t} = \frac{\partial}{\partial x}\left((m-p)[h_\varepsilon(u)]^{m-p-1}\frac{\partial u}{\partial x}\right) + \sin\alpha\frac{\partial}{\partial x}\left([h_\varepsilon(u)]^{m-1}u\right) - E_s u^q \qquad (4)$$

using the regularization kernel $h_\varepsilon(u) = \sqrt{u^2 + \varepsilon^2} \geq \varepsilon$ with a small positive parameter $\varepsilon(= 10^{-10})$, which mathematically as well as numerically rules out the degeneration of the advection and diffusion terms. Application of another conventional regularization method to (1) remains as an option; however, such methods typically rely on transformations of the solution u to a new dependent variable, which in general cannot numerically realize mass conservation [7].

3   Finite Volume Numerical Model

A numerical method for solving the regularized PME (4), which is referred to as the Dual-Finite Volume Method (DFVM) [10], is briefly explained in this section. The DFVM was originally developed for numerically solving Kolmogorov's forward equations, which are linear and conservative advection-diffusion-decay equations (ADDEs), defined on connected graphs [11]. The DFVM concurrently uses primal and dual computational meshes and evaluates the numerical fluxes based on analytical solutions to local two-point boundary value problems. Mathematical analysis revealed that the DFVM for linear ADDEs is unconditionally stable in both space and time with an appropriate implicit temporal integration algorithm, such as the conventional backward Euler method.

In the DFVM, (4) is regarded as a nonlinear ADDE. An operator-splitting algorithm analogous to that of Li et al. [12] specialized for solving (4) is incorporated into the DFVM, so that numerical instability arising from the nonlinear evaporation term is avoided. The time increment between each successive time steps is denoted by Δt, the differential operators defining the advection and diffusion terms of (4) by P_{AD}, and that for the evaporation term by P_E; namely, the operators P_{AD} and P_E for generic sufficiently regular $\varphi = \varphi(t, x)$ are expressed as

$$P_{AD}\varphi = \frac{\partial}{\partial x}\left((m - p)[h_\varepsilon(\varphi)]^{m-p-1}\frac{\partial \varphi}{\partial x}\right) + \sin \alpha \frac{\partial}{\partial x}\left([h_\varepsilon(\varphi)]^{m-1}\varphi\right) \quad (5)$$

and

$$P_E\varphi = -E_s\varphi^q, \quad (6)$$

respectively. Symbolically, the present numerical method temporally discretizes (4) at each time step as

$$u^{(k+1)} = \exp(0.5\Delta t P_E)\exp(\Delta t P_{AD})\exp(0.5\Delta t P_E)u^{(k)} \quad (7)$$

where $u^{(k)}$ is the solution at the k th time step. Temporal integration in the evaporation sub-steps is performed with local analytical solutions to the temporal ordinary differential equation at each node [12]. Temporal integration in the advection-diffusion sub-step is performed with the conventional backward Euler method and the DFVM spatial discretization, which is coupled with a Picard algorithm for handling the nonlinear advection and diffusion terms. Although the DFVM is theoretically unconditionally stable for linear problems, preliminary computation demonstrated that it is not the case for the extended PME with advection term when Δt is sufficiently large. The reason for this phenomenon would be the loss of ellipticity of the discretized system with large Δt as implied in Pop et al. [9].

The present DFVM can be optionally equipped with an adaptive re-meshing technique based on the Moving Mesh Partial Differential Equation (MMPDE) method for accurately resolving sharp transitions of solution profiles [13]. The MMPDE method solves an additional parabolic PDE that governs nodal positions; this is distinct from the primary PDE to be solved, which in this article is the extended PME. An advantageous point of the MMPDE method is that it does not alter the topology of the mesh, which is not common to the other types of adaptive re-meshing methods. The MMPDE method has successfully been used for numerically approximating solutions to the Hamilton-Jacobi-Bellman equations governing upstream fish migration in 1D river flows [14] and Kolmogorov's forward equations in unbounded domains governing probability density functions of stochastically-excited systems [15].

4 Application to Test Cases

The present DFVM is extensively applied to test cases for its verification of accuracy, stability, and versatility. Comparing performance with other published numerical models is beyond the scope of this article, and will be discussed elsewhere. The test cases considered in this article are (Sect. 4.1) Barenblatt problems, (Sect. 4.2) horizontal infiltration subject to evaporation, (Sect. 4.3) inclined infiltration subject to evaporation, (Sect. 4.4) support divorce problem in a horizontal sheet, and (Sect. 4.5) downward infiltration in a vertical sheet.

4.1 Barenblatt Problems

The Barenblatt solutions are analytical solutions to the conventional PME that do not have advection and evaporation terms [16]. A Barenblatt solution has a symmetrical shape in space. The solution has bounded support and it is not differentiable in the classical sense at the fronts of the support, hence it is a weak solution. The Barenblatt solutions and their variants have been used as benchmark test cases for numerical methods for PMEs [16, 17]. Mathematical analysis of Barenblatt solutions and related exact solutions to the PMEs are reviewed in [18, 19]. A difficulty of approximating the Barenblatt solutions lies in the existence of the sharp fronts where numerical solutions would suffer from spurious oscillations as pointed out in Zhang and Wu [17]. For this case we set parameters $p = 0$, $\alpha = 0$, and $E_s = 0$. The initial time is set as $t = 0.01$ for avoiding the Deltaic singularity at $t = 0.00$. The sheet is identified with the domain $\Omega = (-7, 7)$. Homogenous Dirichlet boundary conditions are specified at the boundaries $x = \pm 7$ for the sake of simplicity. The time increment is set as $\Delta t = 0.01$. The numerical and analytical solutions are compared at the time $t = 1.00$ for $m = 2, 3, \ldots, 8$.

Figure 3(a) compares the numerical and analytical solutions for each m without the MMPDE. The fronts of the analytical solutions become sharper as m increases. The numerical solutions do not successfully approximate the position and shape of the front for $m = 6, 7, 8$, whereas they are accurately captured for smaller m. Figure 3(b) provides an enlarged view of numerical and analytical solutions around the right front for $m = 8$, demonstrating that the numerical solution with the MMPDE can more accurately capture sharp solution profile than that without it for challenging Barenblatt problems.

4.2 Horizontal Infiltration Subject to Evaporation

This second test considers water infiltration from one side ($x = 0$) of a horizontal sheet ($\alpha = 0$ and $p = 0$) in an evaporative environment as theoretically considered in Lockington et al. [6]. Assuming that $u = 1$ at the boundary $x = 0$ and that the length of the sheet is sufficiently long, a straightforward calculation shows that the steady solution to this test case is found as

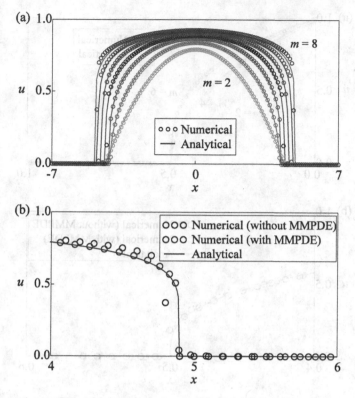

Fig. 3. Comparisons of Barenblatt numerical and analytical solutions (a) over the entire domain and (b) around the right front. (Color figure online)

$$u(x) = \max\left\{\left(1 - (m - q)\sqrt{\frac{E_s}{2m(m+q)}}x\right)^{\frac{2}{m-q}}, 0\right\}. \tag{8}$$

The solution has bounded support for $m > q$. As can be directly seen from (8), this solution is also a weak solution since it is continuous but not differentiable at an edge of the support. The focus here is whether the DFVM can capture the solution profiles and the front position. The sheet is identified with the domain $\Omega = (0, 1)$, which is uniformly discretized into 100 cells. The time increment is set as $\Delta t = 0.001$. Steady numerical solutions are computed starting from the initial guess $u = 0$ over Ω. The parameters E_s, m, and q can be chosen such that the support of (8) becomes $[0, 0.5]$. We have fixed the parameter q at 0.01, and shown solutions for $m = 2, 4$, and 6, so that $E_s = 8.12, 8.06$, and 8.04, respectively. The terminal time of computation is empirically set as $t = 20$ at which the numerical solutions are observed to be sufficiently close to steady state.

Figure 4(a) compares numerical and analytical solutions for different values of m without the MMPDE. The numerical solutions successfully approximate analytical solutions for all m examined. Figure 4(b) provides an enlarged view of numerical and

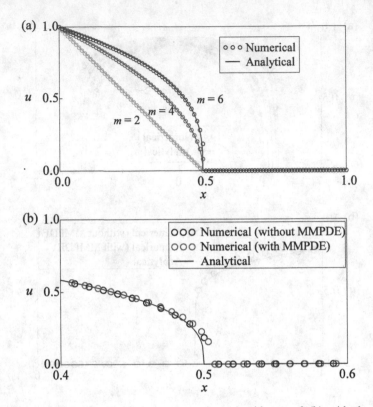

Fig. 4. Horizontal infiltration subject to evaporation (a) without and (b) with the MMPDE. (Color figure online)

analytical solutions around the front for $m = 4$. Numerical solutions without the MMPDE appear to perform better than those with it, indicating that using the MMPDE with the DFVM may not be a good option for the problems with highly singular evaporation term with small q. Qualitatively similar computational results have been obtained for the other values of $m = O(10^0)$.

4.3 Inclined Infiltration Subject to Evaporation

An inclined infiltration problem of water from the downstream-end of a sheet is considered for examining accuracy and stability of the DFVM against more complicated steady problems. As in the previous problem, this test considers water infiltration from one side ($x = 0$) of the sheet. Assuming the relationship $p = q$, we can derive the implicit analytical solution

$$x = -\frac{m \sin \alpha}{E_s} \int_u^1 \frac{v^{m-q-1}}{1 + W_{-1}\left[-\exp\left(-1 - \frac{mv^m \sin^2 \alpha}{(m-q)E_s}\right)\right]} \, dv \tag{9}$$

where $W_{-1}[y]$ denotes the second branch of Lambert W-function, which is valid for the range $\exp(-1) \le y < 1$. To the authors' knowledge, this implicit analytical solution has not been derived elsewhere. The solution (9) can be verified by differentiating twice with respect to x, using the well known result

$$\frac{dW_{-1}[y]}{dy} = \frac{W_{-1}[y]}{y(1 + W_{-1}[y])}. \tag{10}$$

The solution (9) has the bounded support $[0, z]$ where z is the value of the right hand-side of (9) with $u = 0$. The domain Ω is specified as $(0, 2)$. Parameter values are specified as $\alpha = 0.5\pi$ and $p = q = 0.01$, with four cases of (E_s, m) examined; $(3.01, 1)$, $(3.01, 4)$, $(5.01, 1)$, and $(5.01, 4)$, to see the effect that varying nonlinearity of the advection-diffusion terms, and varying strength of the evaporation term have on solution profiles. The domain is uniformly discretized into 200 cells and the time increment is set as $\Delta t = 0.001$. The terminal time of computation is empirically set as $t = 20$ at which the numerical solutions are observed to be sufficiently close to steady state. The MMPDE method is not used since the previous problem showed that it does not to successfully work for the problem with nonlinear evaporation.

Figure 5(a) and (b) compare the numerical and the analytical solution (9) for stronger $(E_s = 1)$ and weaker evaporative environments $(E_s = 4)$. The agreement between two is satisfactory in all cases. The computational results indicate that the DFVM can handle steady infiltration problems in inclined sheets subject to nonlinear evaporation. From the form of the exact solution it follows that the support of each solution is not significantly dependent on the parameter $m = O(10^0)$ if $0 < q \ll 1$, this is demonstrated in Fig. 5.

4.4 Support Divorce Problem in a Horizontal Sheet

The fourth test case is evaporative moisture dynamics in an initially partially wetted horizontal sheet. The values $p = 0$, $\alpha = 0$, and $E_s = 1$ are specified. The sheet is identified with the domain $\Omega = (-1.5\pi, 1.5\pi)$ and the initial condition of u is same as that in Zhang and Wu [17], a continuous function having bounded support with two local extremes and one local minimum. The homogenous Dirichlet condition is assumed at both boundaries of Ω. According to the literature [20, 21], if $m + q = 2$ with $m > 1$ and $q > 0$, the support of the solution separates into two smaller disjoint regions before eventually vanishing. This phenomenon is called "divorce of supports". The remaining model parameters are set as $m = 1.98$ and $q = 0.02$ following Zhang and Wu [17]. Tomoeda et al. [22, 23] mathematically and numerically analyzed similar problems in detail, indicating that the problem focused on in this section does not have classical solutions. The focus here is whether the DFVM can handle this peculiar phenomenon without computational failure. The domain Ω is uniformly divided into 320 cells. The time increment is set as $\Delta t = 0.001$.

Figure 6(a) shows space-time evolution of the numerical solution where its support is identified with the contour line of $u = 10^{-10}$, which is plotted as a black curve. Figure 6(b) plots the spatial solution profile at the time $t = 0.788$ examined in Zhang

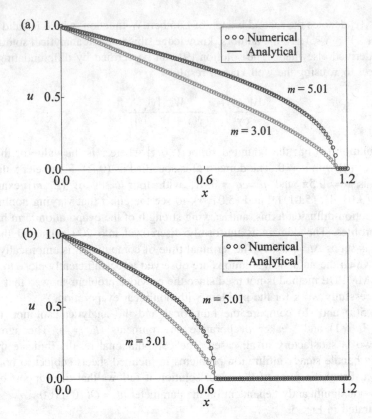

Fig. 5. Comparisons of numerical and analytical solutions for inclined infiltration subject to evaporation for $E_s = 1$ (a) and (b) $E_s = 4$.

and Wu [17]. The obtained results are comparable to those in Zhang and Wu [17] who used a local DG method, which we consider to be a more complicated numerical algorithm than the DFVM because a given computational mesh will have a higher degree of freedom. Hence the computation results show that the DFVM can handle transient problems having an evaporative source term without numerical instability.

4.5 Downward Continuous Infiltration in a Vertical Sheet

The final test case considers a downward water infiltration process from the top of a vertical sheet ($\alpha = -0.5\pi$) subject to an impulsive water input whose analytical solution is available in Hayek [24]. These analytical solutions have been derived in order to comprehend moisture dynamics of wetting front in vertical soil columns with a variety of physical properties. They are therefore not originally considered for moisture dynamics in fibrous sheets, but are used as one of the test cases here because the Richards equation [24] and the extended PME in this paper mathematically have a similar form. The parameters are specified as $p = 1$, the positive integer $m \geq p + 1$, and

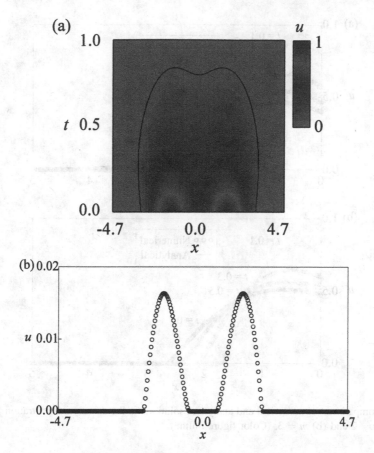

Fig. 6. Numerical solution of the support divorce problem (a) in the $x - t$ phase space and (b) at the time $t = 0.788$. (Color figure online)

$E_s = 0$. The focus here is assessing whether the DFVM can capture infiltration fronts where diffusion and advection terms would degenerate, potentially serving as severe computational difficulties. The sheet is identified with the domain $\Omega = (0,5)$ with the boundary condition $F = 0$ at $x = 0$ and 5 where F is the flux associated with (4). A delta-function initial condition is adopted: $u(0,x) = \delta(x)$. The deltaic condition is approximated as h^{-1} where h is the size of the leftmost cell. The domain Ω is uniformly divided into N regular cells and the time increment is set as N^{-1}. The values of N examined are $100 \cdot 2^k$ for $k = 0, 1, 2, 3, 4$. The MMPDE method is not used here because preliminary computation indicated instability of numerical solutions just after the time $t = 0$, due to the deltaic initial condition.

Figure 7(a) and (b) compare numerical ($N = 200$) and analytical solutions with $m = 2$ and $m = 3$ at selected time steps at which the boundary effects at $x = 5$ are considered to be sufficiently small. The analytical solutions with $m = 2$ and $m = 3$ have unbounded and bounded supports, respectively, and present profiles with different curvatures. They therefore have qualitatively different mathematical properties.

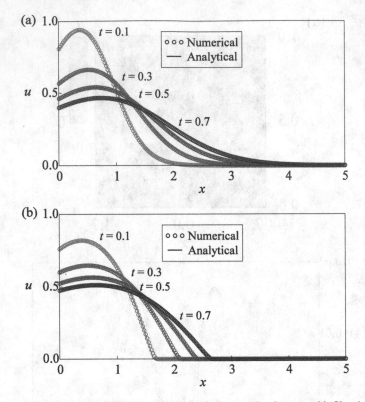

Fig. 7. Comparisons of numerical and analytical solutions to the downward infiltration problem with (a) $m = 2$ and (b) $m = 3$. (Color figure online)

The figures show that the numerical solutions compare well with the analytical solutions where the maximum values and the tails of the solutions are quite accurately reproduced at each time step. The computational results with other values of N, not presented in this article, implied almost first-order convergence of the numerical solutions in both cases. This observation is consistent with the fact that the DFVM is equivalent to a fully-upwind FVM for advection-dominant ADDEs [11].

5 Conclusions

A numerical model of the extended PME with a regularization method for the nonlinear coefficients was proposed and applied to test cases whose solution behaviors are known. The obtained computational results demonstrated its satisfactory accuracy, stability, and versatility, indicating that it can potentially serve as an effective tool for simulating complex moisture dynamics in non-woven fibrous sheets. There were indications that the MMPDE method may not be a good option for problems with highly singular evaporation term with small q. A new exact solution for extended PMEs was derived in this paper, and was accurately reproduced with the DFVM.

The solution may be useful for verification of numerical methods and also for parameter identification of non-woven fibrous sheets and porous media. Spatially 2-D extension of the present mathematical and numerical models does not encounter additional technical difficulties. Future research will address mathematical analysis of the extended PME, on its regularity and analytical solutions in particular following [25]. In addition, assessing consistency error between the original and regularized extended PMEs will be addressed; we expect that the error can be a decreasing function of the parameter ε under an appropriate normed space as in Atkinson and Han [26]. Detailed numerical comparisons of a simplified extended PME with experimental observations have been performed in Yoshioka et al. [1]. Their approach can be further sophisticated with the help of the present numerical model for more realistic and reliable assessment of moisture dynamics in non-woven fibrous sheets. Finally, the presented regularization method and the DFVM can be applied to other degenerate nonlinear diffusion equations arising in applied problems, such as Keller-Segel systems for chemotaxis [27] and population dynamics models [28].

Acknowledgements. We thank Dr. Ichiro Kita and Dr. Kotaro Fukada in Faculty of Life and Environmental Science, Shimane University, Japan for providing helpful comments and suggestions on this article.

References

1. Yoshioka, H., Ito, Y., Kita, I., Fukada, K.: A stable finite volume method for extended porous medium equations and its application to identifying physical properties of a thin non-woven fibrous sheet. In: Proceedings of JSST2015, pp. 398–401 (2015)
2. Landeryou, M., Eames, I., Cottenden, A.: Infiltration into inclined fibrous sheets. J. Fluid Mech. **529**, 173–193 (2005)
3. Wilhelmsson, H.: Simultaneous diffusion and reaction processes in plasma dynamics. Phys. Rev. A **38**, 1482–1489 (1988)
4. Haerns, J., Van Gorder, R.A.: Classical implicit travelling wave solutions for a quasilinear convection-diffusion equation. New Astron. **17**, 705–710 (2012)
5. Broadbridge, P., White, I.: Constant rate rainfall infiltration: a versatile nonlinear model. Analytic solution. Water Resour. Res. **24**, 145–154 (1988)
6. Lockington, D.A., Parlange, J.Y., Lenkopane, M.: Capillary absorption in porous sheets and surfaces subject to evaporation. Transport Porous Med. **68**, 29–36 (2007)
7. Brooks, R.H., Corey, A.T.: Hydraulic properties of porous media. Colorado State University, Hydrology Papers, Fort Collins, Colorado (1964)
8. Stewart, J.M., Broadbridge, P.: Calculation of humidity during evaporation from soil. Adv. Water Resour. **22**, 495–505 (1999)
9. Pop, I.S., Radu, F., Knabner, P.: Mixed finite elements for the Richards' equation: linearization procedure. J. Comput. Appl. Math. **168**, 365–373 (2004)
10. Yoshioka, H.: On dual-finite volume methods for extended porous medium equations, arXiv preprint. arXiv:1507.05281 (2015)
11. Yoshioka, H., Unami, K.: A cell-vertex finite volume scheme for solute transport equations in open channel networks. Prob. Eng. Mech. **31**, 30–38 (2013)

12. Li, Y., Lee, G., Jeong, D., Kim, J.: An unconditionally stable hybrid numerical method for solving the Allen-Cahn equation. Comput. Math Appl. **60**, 1591–1606 (2010)
13. Huang, W., Russel, R.D.: Adaptive Moving Mesh Methods, pp. 27–133. Springer, Heidelberg (2011)
14. Yoshioka, H., Unami, K., Fujihara, M.: A Petrov-Galerkin finite element scheme for 1-D tome-independent Hamilton-Jacobi-Bellman equations. J. JSCE. Ser. A2, **71** (in press)
15. Yaegashi, Y., Yoshioka, H., Unami, K., Fujihara, M.: An adaptive finite volume scheme for Kolmogorov's forward equations in 1-D unbounded domains. J. JSCE. Ser. A2, **71** (in press)
16. Li, H., Farthing, M.W., Dawson, C.N., Miller, C.T.: Local discontinuous Galerkin approximations to Richards' equation. Adv. Water Resour. **30**, 555–575 (2007)
17. Zhang, Q., Wu, Z.L.: Numerical simulation for porous medium equation by local discontinuous Galerkin finite element method. J. Sci. Comput. **38**, 127–148 (2009)
18. Aronson, D.G., Caffarelli, L.A.: The initial trace of a solution of the porous medium equation. Trans. Am. Math. Soc. **280**, 351–366 (1983)
19. Fila, M., Vázquez, J.L., Winkler, M., Yanagida, E.: Rate of convergence to Barenblatt profiles for the fast diffusion equation. Arch. Ration. Mech. An. **204**, 599–625 (2012)
20. Rosenau, P., Kamin, S.: Thermal waves in an absorbing and convecting medium. Physica D **8**, 273–283 (1983)
21. Nakaki, T., Tomoeda, K.: A finite difference scheme for some nonlinear diffusion equations in an absorbing medium: support splitting phenomena. SIAM J. Numer. Anal. **40**, 945–954 (2002)
22. Tomoeda, K.: Numerically repeated support splitting and merging phenomena in a porous media equation with strong absorption. J. Math-for-Ind. **3**, 61–68 (2011)
23. Tomoeda, K.: Numerical and mathematical approach to support splitting and merging phenomena in the behaviour of non-stationary seepage. Theor. Appl. Mech. Jpn. **63**, 15–23 (2015)
24. Hayek, M.: Water pulse migration through semi-infinite vertical unsaturated porous column with special relative-permeability functions: exact solutions. J. Hydrol. **517**, 668–676 (2014)
25. Vazquez, J.L.: The Porous Medium Equation. Oxford University Press, Oxford (2007)
26. Atkinson, K., Han, W.: Theoretical Numerical Analysis, pp. 449–451. Springer, Heidelberg (2009)
27. Bian, S., Liu, J.G.: Dynamic and steady states for multi-dimensional Keller-Segel model with diffusion exponent $m > 0$. Commun. Math. Phys. **323**, 1017–1070 (2013)
28. Naldi, G., Cavalli, F., Perugia, I.: Discontinuous Galerkin approximation of porous Fisher-Kolmogorov equations. Commun. Appl. Indust. Math. **4** (2013). doi:10.1685/journal.caim.446

An Integrated Network Modeling for Road Maps

Zhichao Song[1]([✉]), Kai Sheng[2], Peng Zhang[1], Zhen Li[1], Bin Chen[1], and Xiaogang Qiu[1]

[1] College of Information System and Management, National University of Defense Technology,
Changsha 410073, Hunan, People's Republic of China
song_zhichao@139.com
[2] Electronic Engineering College, Naval University of Engineering, Wuhan 430033, Hubei,
People's Republic of China

Abstract. Critical-location identification on a road map is very helpful to assign traffic resources reasonably in traffic simulations. To simultaneously identify the critical levels of roads and junctions in a road map by the same measure of centrality, we define a novel road network modeling concept: integrated graph, in which both junctions and roads are abstracted as nodes. Based on this method, we analyze the importance of locations in a small road network and Beijing's main-road network. The results show that this modeling method of road networks is feasible and efficient.

Keywords: Network modeling · Road network · Centrality indices · Critical-location identification

1 Introduction

Road network is an important part of environment models in both social simulation and military simulation. Identifying the critical junctions or roads on the whole road map and pay more attention to them is very helpful to assign traffic resources reasonably in traffic simulations. As the traffic resources are always limited, more reasonable the assignment is means that more important parts of the road network can be protected better, so these important parts can be restored very soon once they are invalidated and traffic on the road network can be more stable and more smooth.

Maybe it is because the roads linking with each other can be regarded as a graph or network naturally. The issue that identifying key parts on road maps is usually dealt with by graph theory or complex network theory. Urška et al. [1] researched how to identify critical locations in a spatial network and extracted the key nodes in the street network of Helsinki Metropolitan Area. Porta et al. [2] studied how centrality works in cities based on the road graphs. Wu et al. [3] analyzed the topological bottleneck for traffic networks.

Currently, there are three broadly used road network modeling methods: "Name Street" method [4], primal graph approach [5] and dual graph approach [6]. In the street network modeled by "Name Street" method, nodes represent named streets and edges describe the junctions among these streets. The named-street-centred network reflects a higher level of abstraction topological connections in a given street network and is very

© Springer Science+Business Media Singapore 2016
S.Y. Ohn and S.D. Chi (Eds.): AsiaSim 2015, CCIS 603, pp. 17–27, 2016.
DOI: 10.1007/978-981-10-2158-9_2

suitable to analyze the topological character of the street network. Based on the network, Jiang and Claramunt [4] gave their conclusions that the topology of street networks reveals a small-world property but has no scale-free property. Whereas, this network model does not fit the high-resolution critical-location identification in a road map. On the one hand, the critical-location identification needs to consider geometrical information but lengths of the streets are not described in the model. On the other hand, a named street is usually composed of many road segments and functions of these segments in a whole road network should be different, so the critical location analysis in the network can only give a large-scale location. In the dual graph of a road map, nodes represent roads and edges represent junctions. This graph can provide a road-segment-level critical-location analysis. However the length of each road segment still cannot be considered. The critical-location analysis based on a dual graph is just a topological step-distance concept. The primal graph of a road map is more comprehensive, objective and realistic for the network analysis of centrality. In the graph, nodes are junctions and edges are roads. This graph is usually a weighted graph because lengths of roads are assigned to the edges. Therefore, the primal graph of a road map is the most suitable of the three for the critical-location analysis on a road map.

Centrality indices, which are usually used to measure the importance of the node in a network, are also used to identify the critical levels of the locations on a road map. However, even based on the primal graph method, it still does not satisfy enough to simultaneously calculate the critical levels of roads and junctions of a road network by a measure of centrality once. Mostly, centrality scores of nodes are calculated firstly and then the critical levels of edges are estimated by the scores of nodes they connect directly. For example each edge is assigned a centrality score equal to the average score of the pair of nodes on the both sides of it in literature [5] and several kinds of "edge degree" are defined in literature [7]. In this method, critical levels of edges which are estimated by nodes' scores are totally depended on the nodes on their both sides. It is in such a local scale that actual effect on the whole network an edge has is not considered accurately.

To settle this simultaneous calculation for the centrality of nodes and edges, we propose a new road network modeling method named "Integrated Graph Method" in this paper. In the integrated graph of a road map, junctions and roads are all abstracted as nodes and the connective relations are regarded as edges. The remainder of this paper is organized as follows. In Sect. 2, the integrated graph method is introduced. In Sect. 3, we present the key parts identification for a small road map. In Sect. 4, a real large-scale case is given. Our conclusions are given in Sect. 5.

2   Method

We consider that no matter it is a road or a junction the critical locations on the road map should be identified by one measure of centrality once. As we have introduced, the road network modeling methods such as primal graph method, dual graph method or "Named Street" method cannot meet our demands. Among the three methods, primal graph is much more suitable for discovering critical elements of the network. However,

in primal graph network the importance of nodes and edges still cannot be simultaneously calculated by one measure of centrality. Always the edges' importance is estimated
by the centrality sores of nodes connecting the edge directly. In this paper, we integrate
the ideas of node presentation in primal graph and dual graph that both roads and junctions on a road map are represented by nodes. That is why the network modeling method
we proposed is named "Integrated Graph Method".

Specifically, both roads and junctions on a road map are represented by nodes when
the road network is constructed. If a road is directly connected with a junction, there is
an edge between them and the weight of every edge is half of the road's length. Based
on this weighted graph or network, the importance of roads and junctions can be
computed simultaneously by one measure of centrality. Figure 1 gives the different

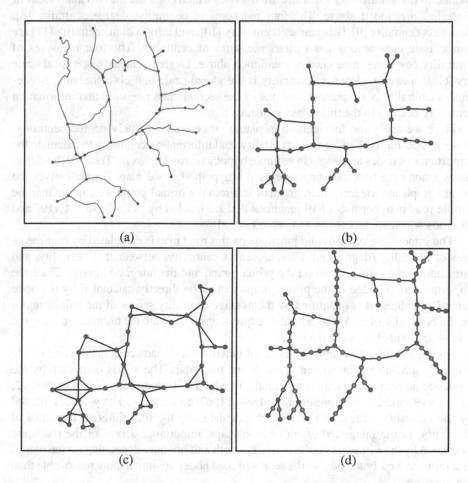

(a)   (b)

(c)   (d)

Fig. 1. The different network models of a simple road map: (a) is the original road map; (b) is
the primal graph of the road map, in which nodes are junctions and lines are roads; (c) is the dual
graph of the road map, in which nodes mean roads and lines mean junctions; (d) is the network
modeled by integrated graph method, in which red nodes represent junctions and green nodes
represent roads (Color figure online)

network models of a simple road map based on primal graph method, dual graph method and the integrated graph method.

3   Key Parts Identification for a Small Road Map

Being the importance measure methods, centrality indices are often used to identify key parts on road maps. So far, there have been various measures of structural centrality, which can be divided into three classes. The first-class measures are based on the idea that the a central node in a network is near to the other nodes. The second-class measures are based on the idea that central nodes stand between other node pairs on their paths of connection. The third-class measure are the ones which combine the two main ideas of centrality mentioned above. The four measures of centrality: degree centrality [8], closeness centrality [9], betweenness centrality [10], and information centrality [11] are four classic ones among the various measures of centrality. The four measures of centrality cover the three classes mentioned above. Degree centrality is a local-scale first-class measure, closeness centrality is the global-scale first-class measure, betweenness centrality is a representative one of the second-class measure, and information centrality belongs to the third-class measure.

Here we apply the four classic measures of centrality namely degree centrality, closeness centrality, betweenness centrality, and information centrality to calculate the importance of nodes and edges in a simple hypothetic road network. To show the differences among the four measures, we used a hypothetic road map. Figure 2 gives the primal graph and the integrated graph of it. From the primal graph we can see that the simple road map consists of 19 junctions that are labeled by "v1, v2, v3 ..., v19" and 18 roads which are labeled by "e1, e2, e3 ..., e18".

The importance of roads and junctions on this road map is calculated by four measures of centrality (degree centrality, closeness centrality, betweenness centrality, and information centrality) based on the primal graph and the integrated graph. Since the importance of an edge in the primal graph cannot be directly calculated by the node centrality indices, it is quantified by the average centrality scores of the pair of nodes on the both sides of it. As the centrality expressions we use are the normalized ones, the results are limited between 0 and 1.

Figure 3 shows the centrality scores of junctions and roads calculated based on the primal graph and the integrated graph of the road map. The x-axis is marked by the sequence numbers of junctions and roads, in which 1–19 represent the nodes "v1, v2, v3 ..., v19" and 20–37 represent the edges "e1, e2, e3 ..., e18". The y-axis is marked by the centrality scores. The scores are calculated by the four different measures of centrality. In the integrated graph of road maps, importance scores of the roads are evaluated by their impacts on the whole network and not only depending on the pair of junctions on their both ends, so the scores of road nodes are much more reasonable than the averages.

The centrality scores calculated in the integrated graph are universally lower than those in the primal graph except the closeness. This is because the primal graph does not consider the roads as nodes and the normalization factor is related to the node

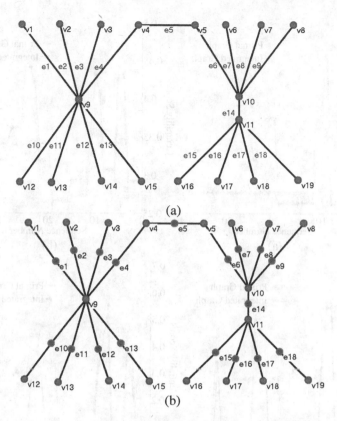

Fig. 2. A simple hypothetic road network: (a) is the primal graph and (b) is the integrated graph

number. If the normalization factors applied to the two network graphs are the same, the degree centrality sores of junctions in the two networks will be the same. Whereas the importance of roads are not exact when degree centrality is used in the integrated graph. We should not ascribe this to the road network model, and it is because the shortage of degree centrality itself that degree centrality can only give a so local scale evaluated result. For example, the importance of "v4" and "v5", of which scores are very high when closeness centrality, betweenness centrality and information centrality are used, is not very obvious in Fig. 3(a). For the other three measures of centrality, effects come from the normalization factor are not obvious. It is counteracted by the decrease of node pairs' distances for closeness centrality and the increase of the shortest path for betweenness centrality. Moreover, the increase of the shortest path makes an extra contribution to heighten the closeness scores. The normalization factor of information centrality is actually not affected by the node number. The decrease of information centrality scores in integrated graph is only due to the increase of network efficiency caused by the new road nodes. The trend of the four centrality scores in the two graphs is similar and the consistency of the closeness centrality scores is the best. The relative importance of junctions does not change very much in the integrated graph. Especially when betweenness centrality is used, importance of junctions in these two networks is

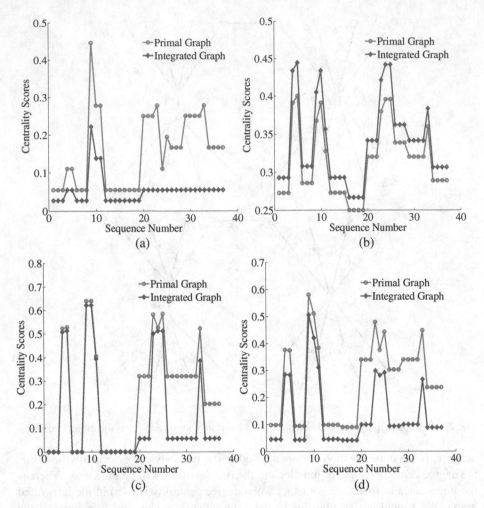

Fig. 3. Importance of junctions and roads calculated based on the two road networks by different measures of centrality: (a) is calculated by degree centrality; (b) is calculated by closeness centrality; (c) is calculated by betweenness centrality; (d) is calculated by information centrality (Color figure online)

nearly the same (Fig. 3(c)). That is because the measure of betweenness centrality is the least affected by the new nodes in integrated graphs.

From the comparison of the roads' scores in the two network graphs, we can conclude that the evaluation of junctions' importance is hardly affected by the road nodes and the calculation of road's importance can be more exact when integrated graph method is used, because the scores are calculated based on the contributions of the road nodes in the whole network.

4 Real Large-Scale Case: An Analysis of Beijing's Main Road Network

In this section, we will use the integrated graph method to model the Beijing's main-road map which is a large-scale road network and then simultaneously compute the importance of junctions and road segments by the three classic global-scale centrality namely closeness centrality, betweenness centrality and information centrality. As shown in Fig. 4, there are 4192 junctions and 5166 road segments in the map. The line only depicts the shape, location and length of a main road between two junctions. As we have discussed, the road network should be modeled using the integrated graph method before the importance of roads and junctions is computed. The network which was built using the integrated graph method has 9358 vertices and 10332 edges and is shown in Fig. 5, in which red vertices represent junctions and green vertices are roads.

Fig. 4. The main road map of Beijing

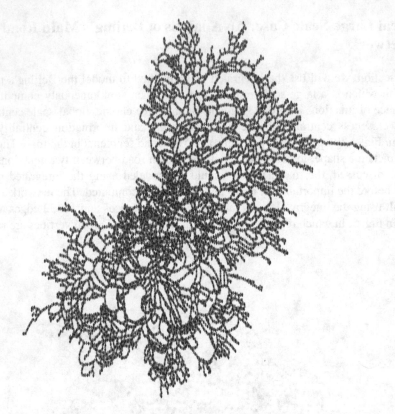

Fig. 5. The integrated graph of the main road map of Beijing (Color figure online)

Based on the integrated network model, the importance of roads and junctions are computed by closeness centrality, betweenness centrality and information centrality. Degree centrality is not considered here, because it is a local-scale evaluated method and is not suitable to evaluate the importance of road nodes in the integrated graph. Actually, when the information centrality is used, the junctions' importance is evaluated by the group information centrality [11]. Because we consider that if a junction is invalidated the roads that directly connect it are always useless too.

The cumulative distributions of the three measures of centrality are shown in Fig. 6. The cumulative distribution probability is defined by $P(C) = \int_{C\infty} N(C')/N \, dC'$, where $N(C)$ is the number of nodes whose centrality scores equal to C and N is the number of nodes in the whole network. The centrality scores in the figure are all renormalized by dividing the maximum ones, so the maximum scores are 1. From it we can have conclusions as follows:

- Closeness centrality is mainly linear and betweenness centrality shows an obvious exponential distribution. Information centrality which is the third-class centrality measure also has an exponential character.
- Importance levels computed by closeness centrality are more uniform than the other two but they do not cover the whole range 0–1. Betweenness centrality and

information centrality have a better resolving power for the nodes' importance and are both cover the whole range.

- We can see from the figure that the distribution curve of information centrality scores locates between the closeness centrality and betweenness centrality. Maybe it is because information centrality is the third-class measure of centrality which combines the two ideas of closeness centrality and betweenness centrality.

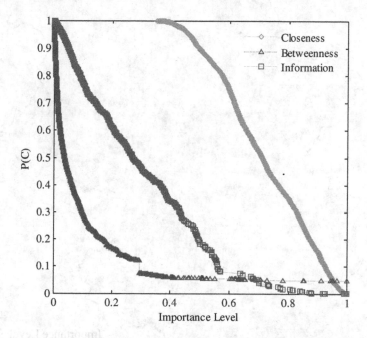

Fig. 6. The cumulative distributions of the three classic global-scale measures of centrality in the integrated graph of Beijing's main road map. (Color figure online)

Considering information centrality combines the two ideas of closeness centrality and betweenness centrality, we think it should be one comprehensive method to measure the importance of junctions and roads in a map, so we have mapped importance of junctions and roads computed by information centrality in the map and show the geographic distribution of them in Fig. 7. In the figure, the centrality scores are renormalized by dividing the maximum one and the nodes and edges are marked by a series of gradual changed colors (as shown at the right bottom of Fig. 7). From the result map, we found that many important junctions and roads we identified are the real high-grade roads. As road design in a city is always very scientific and reasonable, this phenomenon shows the validity of our road network modeling method.

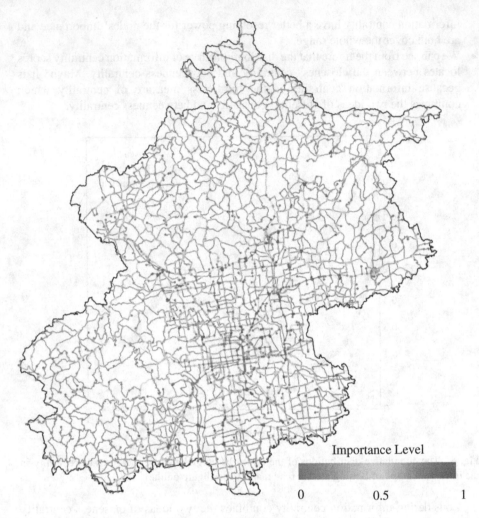

Fig. 7. The geographic distribution of critical roads and junctions in Beijing's main road map (Color figure online)

5   Conclusions

This paper proposes a road network modeling method called integrated graph to settle the simultaneous computation of the importance of roads and junctions on a road map in traffic simulations. This method that both roads and junctions on a road map are represented by nodes integrates the ideas of node presentation in primal graph and dual graph. Based on this, the importance of junctions and roads can be evaluated by a measure of centrality once.

The method has been validated by analyzing the importance of locations in a small road network and Beijing's main road network. The results show that our modeling method of road networks is feasible and efficient. The calculation of a road's importance in the integrated network is based on its contribution to the whole network, and is more reasonable than the evaluation based on the junctions on its both sides. Furthermore, the integrated network modeling method makes the importance of roads and junctions measured by the same standard, so the importance of roads and junctions can be compared more accurately and clearly. These improvements are very helpful to identify the critical junctions or roads on the whole road map in traffic simulations and realize a more reasonable traffic resources assignment in traffic simulations.

Acknowledgements. The authors would like to thank National Nature and Science Foundation of China under Grant Nos. 91024030, 91224008, 61503402, and 71303252.

References

1. Demšar, U., Špatenková, O., Virrantaus, K.: Identifying critical locations in a spatial network with graph theory. Trans. GIS **12**, 61–82 (2008)
2. Porta, S., Latora, V.: Centrality and cities: multiple centrality assessment as a tool for urban analysis and design. In: New Urbanism and Beyond: Designing Cities for the Future, pp. 140–145 (2008)
3. Wu, J., Gao, Z., Sun, H.: Topological-based bottleneck analysis and improvement strategies for traffic networks. Sci. China Ser. E Technol. Sci. **52**, 2814–2822 (2009)
4. Jiang, B., Claramunt, C.: Topological analysis of urban street networks. Environ. Plann. B **31**, 151–162 (2004)
5. Porta, S., Crucitti, P., Latora, V.: The network analysis of urban streets: a primal approach. Environ. Plann. B Plann. Des. **33**, 705–725 (2005)
6. Porta, S., Crucitti, P., Latora, V.: The network analysis of urban streets: a dual approach. Phys. A Stat. Mech. Appl. **369**, 853–866 (2006)
7. Nieminen, J.: On the centrality in a graph. Scand. J. Psychol. **15**, 332–336 (1974)
8. Holme, P., Kim, B.J.: Attack vulnerability of complex networks. Phys. Rev. E **65**, 056109 (2002)
9. Latora, V., Marchiori, M.: Efficient behavior of small-world networks. Phys. Rev. Lett. **87**, 198701 (2001)
10. Freeman, L.C.: Centrality in social networks conceptual clarification. Soc. Netw. **1**, 215–239 (1979)
11. Latora, V., Marchiori, M.: A measure of centrality based on network efficiency. New J. Phys. **9**, 188 (2007)

A Novel Flexible Experiment Design Method

Gang Zhai, Yaofei Ma[✉], Xiao Song, and Yulin Wu

College of Automation Science and Electrical Engineering, Beihang University,
Beijing 100083, China
mayaofeibuaa@163.com

Abstract. Latin Hypercube Design (LHD) is a traditional method of Design Of Experiments (DOE) and is often employed in system analysis. However, this method imposes restriction on experiment trials and needs much computation capacity to obtain the optimal design. A novel experiment design method called ETPLHD is proposed in this paper to solve this problem. ETPLHD can control the number of design points and thus presents more flexibility to control the number of experiment trails, which is more efficient compared to the fixed experiment trails in the traditional LHD method for a same design space. An experiment was conducted to compare ETPLHD with the other two experiment design algorithms. The results showed that TPLHD reveals high design performance and less time consumption.

Keywords: Experiment design · Simulation · Latin Hypercube Design

1 Introduction

Simulation is a promising way to study the complex systems with high performance-price ratio [1]. To analysis the characteristics of the target system, normally a large number of experiment trails need to be run and different level combinations of the parameters are tested. While the high-performance computation facilities are widely used to speed up the computation, the computation capacity required in system analysis still makes it difficult to test each level combination of all parameters, especial for those complex systems with large set of parameters. Even a simple application with 5 parameters, and each parameter contains 10 levels, 10^6 min (over 2 years) is required to test all level combinations assuming each trail need 10 min to run.

As a result, the DOE technique is widely used in any experiment-based domains [2] for its capability to analyze the target system with less experiment trials. The DOE method normally selects a small amount of typical experiment points in the parameter space, to obtain the comprehensive understanding of the target system. It is not necessary to test each level combination of the parameters, thus improving the experiment efficiency greatly. The model of the target system is expected to be derived based on this small amount of experiments, then the further analysis even the prediction can be made.

© Springer Science+Business Media Singapore 2016
S.Y. Ohn and S.D. Chi (Eds.): AsiaSim 2015, CCIS 603, pp. 28–39, 2016.
DOI: 10.1007/978-981-10-2158-9_3

A critical property is the space-filling property, i.e., how the design points distributed in the experiment space. Among various DOE methods, the Latin Hypercube Design (LHD), which was proposed by McKay [3], is most used in simulations [4]. It is always the concern in different variants of LHD design to obtain better space-filling property. Park developed a row-wise element exchange algorithm to obtain the optimal LHD [5]. Morris and Mitchell applied simulated annealing algorithm to optimize the design [6]. Bates et al. employed the Genetic algorithm to optimize LHD [7]. Although these variants present good performance, the involved number of experiment trails is large. Felipe Viana et al. proposed a fast optimal Latin Hypercube Design using Translational Propagation algorithm (TPLHD) [8]. TPLHD generates nearly optimal design instead of the global optimal design. However, the number of experiment trials of TPLHD is fixed for discrete experiment space, which is often insufficient to fully explore the experiment space.

In this paper, an Extended TPLHD method (ETPLHD) is proposed to generate the design points flexibly. ETPLHD can generate more points in the design space than traditional LHDs, which will be helpful in study of the target system. It was shown in the comparison with other DOE methods that ETPLHD is efficient to produce the experiment points, and achieved better evaluation results about the studied system.

The rest of this paper is organized as follows. Section 2 gives a brief introduction of TPLHD. Then the proposed method ETPLHD is described in detail. Section 3 introduces the linear regression model for system approximation and predication. Section 3.2 presents the comparative experiment between ETPLHD and the other two DOE methods, and the results analysis is given. Finally, the remarkable features about ETPLHD approach is concluded in Sect. 4.

2 Method

2.1 Review of Latin Hypercube via Translational Propagation (TPLHD)

TPLHD method works in real time at the cost of finding the near optimal design instead of the globally optimal design. Suppose a design space with n_v variables, each variable has n_p levels. The first step in TPLHD is to create a seed design that contains n_s points. This seed design is used as a pattern to fill the experiment space iteratively. To fill the space, the experiment space is partitioned into n_b blocks firstly:

$$n_b = n_p/n_s \tag{1}$$

The number of levels contained by each block is determined as follows:

$$n_d = (n_b)^{1/n_v} \tag{2}$$

The second step is to fill the seed design into each block. An example is illustrated in Fig. 1. A seed design containing only 1 point is created in a 9×2

(two variables and each has 9 levels) design space. Then the original space is divided into $9/1 = 9$ blocks, and each block has $9^{1/2} = 3$ levels. As Fig. 1(a) shows, the seed design is placed in the left-bottom block firstly. Then the seed design is iteratively shifted by $n_p/n_d = 3$ levels along one dimension until this dimension is filled with the seed design, as Fig. 1(b) and (c) shows. Next, the design along this dimension is adopted as a new seed design to fill along other dimension until all blocks are filled, as shown in Fig. 1(d).

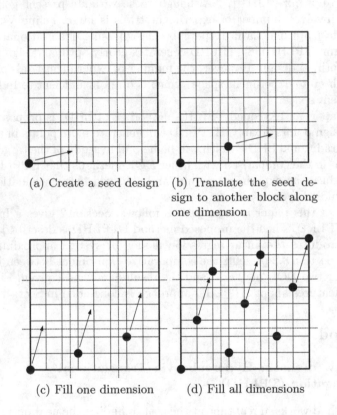

(a) Create a seed design

(b) Translate the seed design to another block along one dimension

(c) Fill one dimension

(d) Fill all dimensions

Fig. 1. The procedure of creating 9×2 TPLHD

As all the key operations are to translate the design points, The TPLHD method requires much less computation compared to other optimal methods which normally need to search the best design among all $(n_p!)^{n_v}$ LHD designs.

A criterion parameter ϕ_p, is used to measure the space-filling quality of the experiment design. For an experiment space, a smaller ϕ_p indicates that the created experiment points are better in distribution to fill up the space.

$$\phi_p = [\sum_{i=1}^{n_p-1} \sum_{j=i+1}^{n_p} d_{ij}^{-p}]^{1/p} \qquad (3)$$

where p is a pre-selected integer value and d_{ij} is the distance between any two design points x_i, x_j:

$$d_{ij} = d(x_i, x_j) = [\sum_{k=1}^{n_v} |x_{ik} - x_{jk}|^t]^{1/t} \tag{4}$$

ϕ_p is also adopted in this paper to measure the space-filling quality of experiment design. As suggested by Jin et al. [9], the value $p = 50, t = 1$ is taken here. TPLHD method works well with experiment design less than 6 variables. It is difficult to approximate a good experiment design in high-dimensional space for 2 reasons: (i) the Curse of Dimensionality. The partitioned blocks in experiment space will grow exponentially with dimensions; (ii) the distribution of experiment points will be asymmetry in high-dimensional space with linear partition.

2.2   The Extented TPLHD (ETPLHD)

The number of experiment points in TPLHD is constrained by n_p, which is the levels of the variable. This characteristic will lead to a small point set most of the time. For example, a size of 10×5 experiment design space contains total 5^{10} level combinations, however, only 10 experiment points would be chosen with TPLHD method. Obviously, this small amount of experiments is insufficient to analysis the target system.

To increase the experiment points, the ETPLHD s proposed in this paper to provide a layered design approach to obtain better distribution of the experiment points. The horizontal interpolation is employed in ETPLHD to design the experiment in an expanded space and then is scaled into the origin space.

Assuming all variables having the same number of levels, Eq. (2) is modified as:

$$d = n_p^{1/2} \tag{5}$$

d is expected to be an integer value, thus n_p needs to be rounded as the value that can be squared such as 4, 9 or 16. For instance, $n_p = 15$ can be rounded up to 16, i.e. one extra level is added into the experiment space, then the points corresponding to the extra level are eliminated at last to correct the design.

ETPLHD holds that each block contains the same number of points as TPLHD. However, the number of blocks in ETPLHD is much bigger than that in TPLHD for cases with more than two dimensions, leading to the increment of the design points in ETPLHD (denoted as n^*).

$$n^* = d^{n_v} = n_p^{n_v/2} \tag{6}$$

Consider a $n \times m$ (n levels, m variables) experiment space. Initially, one dimension (variable) is chosen to be divided into intervals according to Eq. (5). After this operation, the experiment design is conducted in a subspace $n \times (m-1)$ within each block, where the dimension that has been divided is excluded. The design in subspace $n \times (m-1)$ takes the same steps:chose one dimension to divide,

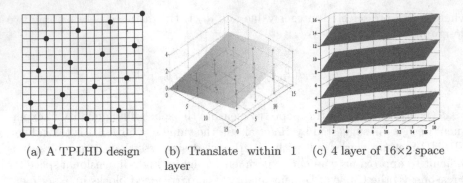

(a) A TPLHD design   (b) Translate within 1   (c) 4 layer of 16×2 space
                        layer

Fig. 2. The procedure to create 16 × 3 ETPLHD

and then perform the experiment design in a lower dimensional subspace, until
the TPLHD method is applied.

Figure 2 illustrates this process taking a design in 16 × 3 experiment space.
First, one dimension (denoted as the 1^d) of the experiment space is chosen and
partitioned into 4 parts according to Eq. (5). 4 intervals along this dimension are
determined, as shown in Fig. 2(a). Next, a 16 × 2 TPLHD designs (the dimensions are denoted as 2^d and 3^d) is conducted in each interval. By this way, the
ETPLHD generates 64 points. For comparison, TPLHD obtains a design with
16 points.

In this example, the experiment points within low-dimensional subspace need
to be expanded to the higher dimension. Generally, if the points set x_{im} has been
obtained within a subspace of m dimensions, the design points can be expanded
to the higher dimension as follows:

$$x^k_{i(m+1)} = x_{im} mod\ d + (k-1) \cdot d \tag{7}$$

where $k(1 \le k \le d_{m+1})$ is the index of the interval along dimension $m + 1$. The
expansion is repeated until all dimensions are included. As for the example in
Fig. 2, the experiment points form the slanted layer along the dimension 1^d, as
shown in Fig. 2(c).

After the expansion to higher dimension, some points may stay in a small
region of the experiment space. Consider the example in Fig. 2, an experiment point $[x_0, y_0]$ designed in the $2^d \times 3^d$ subspace will generates a series
experiment points along dimension 1^d by taking the expansion operation:
$[x_0, y_0, z_0], [x_0, y_0, z_3], [x_0, y_0, z_7], [x_0, y_0, z_{11}]$. From the direction of the dimension 1^d, they are lines. This problem will be worse when the dimension grows.

As a result, it is necessary to adjust the distribution of the experiment points
before expansion to the higher dimension. Assuming the current dimension is m,
then the points are shifted $p(0 \le p \le d_m)$ unit along the $2 \sim m$ dimension, where
d_m is the number of the intervals along dimension m. Equation (7) is changed as
follows:

$$x^k_{i(m+1)} = (x_{im} + p)mod\ d + (k-1) \cdot d \tag{8}$$

There are two issues should be noted. First, the value of p is normally determined empirically. Second, a control parameter s can be defined to specify how many dimensions need to apply the ETPLHD method. For a $n \times m$ space, $s = 2$ means only two dimensions apply the ETPLHD method and the rest $m-2$ dimensions will apply the TPLHD method. By the control parameters, the number of experiment points determined by ETPLHD is:

$$n^* = n_p \cdot d^s = n_p^{\frac{s}{2}+1} \tag{9}$$

Figure 3 demonstrates the shift within the $2^d \times 3^d$ subspace before expansion to dimension 1^d.

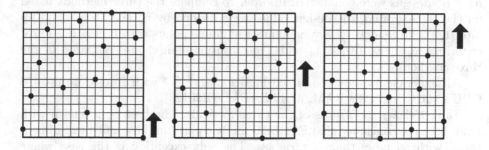

Fig. 3. Translate the seed on each layer

The pseudo code of ETPLHD is shown as follows:

Algorithm 1: $ArrayCreateETPLHD(m, n, s)$

```
var
  m: the number of the design variables (dimensions) of the
     experiment space. The first dim refer to sub-population.
  n: the levels of each variable.
  s: the control variable specifying how many dimensions should
     be applied the ETPLHD design.
begin
  d=sqrt(n);
  Array seed = CreateTPLHD(m-s,n); //Assuming avaiable
  Array res;
  for(dimension=m-s+1:m){
    for(layer=1:d){
      Add the seed design into res;
      Translate the seed by current dimension;
    }
    seed=res;
  }
  return res;
end.
```

2.3   DOE Methods Comparison

Three DOE methods, ETPLHD, TPLHD and a random-selection based LHD method *lhsdesign* (the implemented function name in Matlab) are compared to demonstrate the effectiveness of ETPLHD. The *lhsdesign* method selects the LHD according to the best max-min criterion, i.e., the maximization of the minimum distance d_{min}, from 200 random LHDs. d_{min} is the other measure of the points density (similar with ϕ_p); a bigger d_{min} indicates better distribution of the points.

In a $n_v \times n_p$ (n_v levels, n_p variables) design space, the expected experiments trials n^* can be determined by Eq. (9). With TPLHD and $Matlab^{TM}$ method, $n_v \times n^*$ designs were conducted. In order to compare the three methods based on the same ground, the design values were divided by n_d^s and then round up into $[1, n_p]$, to generate the same number of design points with ETPLHD. Three criterions ϕ_p in Eq. (3), d_{min} and *Time* consumption(s) are compared in the three methods.

All experiments were conducted in $Windows^{TM}$ 7 with Intel Core i5-3470 CPU (3.20 GHz), 4 GB RAM, $MATLAB^{TM}$ (R2012a).

Table 1 shows the results of the three methods. The best ϕ_p is shown with bold. It can be concluded that ETPLHD method performs best under experiment space with no more than 5 variables. The only exception is the case where $n_p = 4, n_v = 4$ and $s = 1$. For the design space with 6 variables, TPLHD performs worse than *lshdesign* with most cases, but ETPLHD is superior to *lshdesign* basically. In addition, ETPLHD can generate points with least time consumption among the three methods. For all cases, ETPLHD and TPLHD cost less time to obtain an optimal design. By contrast, *lhsdesign* requires seconds to minutes for a large experiment space.

3   Apply ETPHLD in System Predication

3.1   Linear Regression Prediction Model

The goal of DOE method is to evaluate the target system with less experiment trails. Many approximation methods such as the linear regression, Kriging model, neural nets, support vector regression and so on are often employed to evaluate and then predict the target system. The linear regression is the most used approach among *them*[10].

A first-order polynomial can be given by

$$y = X\beta + e \tag{10}$$

where $y = (y_1, \ldots, y_n)'$ is the predicted value of the target system with n experiments. $X = (x_{ij})(i = 1, \ldots, n, j = 1, \ldots, q)$ is the experiment data recorded in n trails, where i in the index of experiment trail, and j is the index of data within each trail. $\beta = (\beta_1, \ldots, \beta_q)$ is the regression coefficients; and $e = (e_1, \ldots, e_n)'$ denotes the residuals in each experiment.

Table 1. Performance comparison among ETPLHD, TPLHD and *lshdesign*

			ETPLHD			TPLHD			*lshdesign*		
n_p	n_v	s	ϕ_p	d_{min}	Time	ϕ_p	d_{min}	Time	ϕ_p	d_{min}	Time
3	4	1	2	0.5	≈0	2.056	0.5	≈0	2.027	0.5	0.1
	9	1	2.405	0.444	≈0	3.178	0.333	≈0	4.6	0.222	0.1
	16	1	3.485	0.312	≈0	4.322	0.25	≈0	8.291	0.125	0.3
	25	1	3.485	0.312	≈0	4.322	0.25	≈0	8.291	0.125	0.3
4	4	1	1.333	0.75	≈0	1.014	1	≈0	1.351	0.75	0.1
		2	2.065	0.5	≈0	2.114	0.5	≈0	2.065	0.5	0.1
	9	1	2.281	0.444	≈0	2.313	0.444	≈0	3	0.333	0.1
		2	3.237	0.333	≈0	3.294	0.333	≈0	9.199	0.111	0.43
	16	1	2.704	0.375	≈0	3.304	0.312	≈0	5.333	0.187	0.3
		2	4.439	0.25	≈0	4.505	0.25	≈0	16.35	0.062	3.2
	25	1	4.283	0.24	≈0	5.000	0.2	≈0	6.337	0.16	0.85
		2	5.664	0.2	≈0	5.740	0.2	≈0	25	0.04	18.2
5	4	1	1.014	1	≈0	1.333	0.75	≈0	1.333	0.75	0.1
		2	2.027	0.5	≈0	2.056	0.5	≈0	2.027	0.5	0.1
	9	1	1.521	0.667	≈0	2.281	0.444	≈0	1.8	0.555	0.1
		2	3.174	0.333	≈0	4.626	0.222	≈0	4.562	0.222	0.4
	16	1	2.073	0.5	≈0	3.2	0.312	≈0	2.286	0.437	0.3
		2	4.055	0.25	≈0	4.222	0.25	≈0	4.169	0.25	3.2
	25	1	2.276	0.44	≈0	3.671	0.28	≈0	3.126	0.32	0.9
		2	4.285	0.24	≈0	5.348	0.2	≈0	6.425	0.16	18.1
6	4	1	0.681	1.5	≈0	0.8	1.25	≈0	0.681	1.5	0.1
		2	2	0.5	≈0	2	0.5	≈0	1.333	0.75	0.1
		3	2	0.5	≈0	2.108	0.5	≈0	2.056	0.5	0.15
	9	1	1.045	1	≈0	1.533	0.666	≈0	1.125	0.888	0.15
		2	2.383	0.444	≈0	2.388	0.444	≈0	2.281	0.444	0.45
		3	3.258	0.333	≈0	4.767	0.222	≈0	4.562	0.222	3
	16	1	1.380	0.75	≈0	2.173	0.5	≈0	1.623	0.625	0.3
		2	3.468	0.312	≈0	4.222	0.25	≈0	3.271	0.312	3.3
		3	4.203	0.25	≈0	8.224	0.125	0.3	5.527	0.187	49
	25	1	1.701	0.64	≈0	3.571	0.28	≈0	1.927	0.52	0.9
		2	3.966	0.28	≈0	3.488	0.32	≈0	4.166	0.24	18
		3	4.697	0.24	≈0	5.520	0.2	0.2	8.605	0.12	443.9

Let $w = (w_1, \ldots, w_n)'$ be the true output of the simulation system, the Sum of Squared Residuals(SSR) is given by Eq. (11)

$$SSR = (y - w)'(y - w) \tag{11}$$

The coefficient vector β is defined as follows to compute the predication:

$$\beta = (X'X)^{-1}X'w \tag{12}$$

The accuracy of the model is normally measured by the mean square error (MSE), which is the accurate estimate of the true error of the prediction model.

$$MSE = SSR/n \tag{13}$$

where n is the number of experiment trials.

3.2   A Complex Example: The Combat Simulation

There are blue and red force in the combat scenario. The blue force includes a formation of fighters, who attempts to cross a strait to attack the ships of the red. On the other hand, the red ships have the anti-air capacity. Once the blue fighters flight close to the ships, the SAM missiles will be launched to protect the ships (Fig. 4). The relationship between the performance of SAM missile and the shoot-down number of blue fighters is concerned. The combat is simulated using a Computer Generated Force (CGF) platform [10].

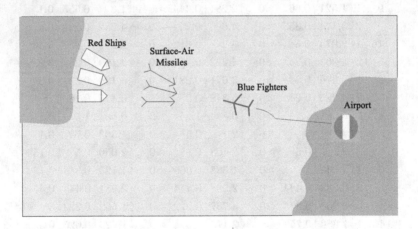

Fig. 4. The two dimension display of the scenario

Five parameters of SAM missile are adopted as the design variables and each has 4 levels, as shown in Table 2.

TPLHD and ETPLHD are tested in this scenario for comparison. In ETPLHD, $s = 3$ produces 32 experiment points. Each experiment point is tested for 5 times to get the mean value of the output for further analysis. The linear regression approach is employed to obtain an approximated system model that can be used to predict the output of the combat. Table 3 demonstrates a part of the results and the predictions of the approximated models based on the two DOE methods.

Table 2. Design parameters and levels

Variables (of the missiles)	Levels			
	1	2	3	4
Velocity X_1 (km/h)	700	800	900	1000
Kill Radius X_2 (m)	25	50	75	100
Range X_3 (km)	250	500	750	1000
Bomb Loads X_4	4	6	8	10
Killing Probability X_5	0.4	0.5	0.6	0.7

Table 3. The true results and the predictions

(a)TPLHD							
No	X_1	X_2	X_3	X_4	X_5	y	y_{pred}
1	1	1	1	1	1	8.2	10.2520
2	3	1	1	3	2	17.2	18.1212
3	4	2	2	2	3	14.4	13.6219
4	2	2	4	4	4	27	28.7731
. . . . ..							
32	4	4	4	4	4	26.2	25.2967
(b)ETPLHD							
No	X_1	X_2	X_3	X_4	X_5	y	y_{pred}
1	3	2	1	1	1	8	8.2456
2	3	3	4	2	1	15.6	17.7121
3	2	1	1	3	2	20.4	19.7148
4	4	1	3	2	3	17.2	16.0479
. . . . ..							
32	4	3	1	1	3	8	6.7885

Given the data in Table 3, the *MSE* of the two methods can be calculated according to Eq. (13). The *MSE* of ETPLHD is 0.6141, which is much less than that (1.0533) of TPLHD. This result indicates that the approximated system model using the ETPLHD method performs better than the model using TPLHD method.

20 random points are selected from the experiment space to verify the effectiveness of the approximated models. Table 4 demonstrates a part of the results. y_a is the prediction by the approximated model that uses the TPLHD method, while y_b is the predication by the model that uses the ETPLHD method.

The plots of the simulation values and two prediction values are shown in Fig. 5. The *MSE* of the two predication is 4.0147 and 3.217 respectively, indicating that the prediction model with ETPLHD is more precise.

Table 4. Part of results of 20 random points with the two methods

No	X_1	X_2	X_3	X_4	X_5	y	y_a	y_b
1	3	1	3	3	4	24.2	22.1132	22.0739
2	2	3	2	2	1	17	14.1160	15.4181
3	4	3	1	2	1	13.2	10.4750	11.6806
4	1	3	1	3	1	21.2	18.4052	19.7464
. . . . . .
20	4	4	2	1	2	8	7.0334	7.9351

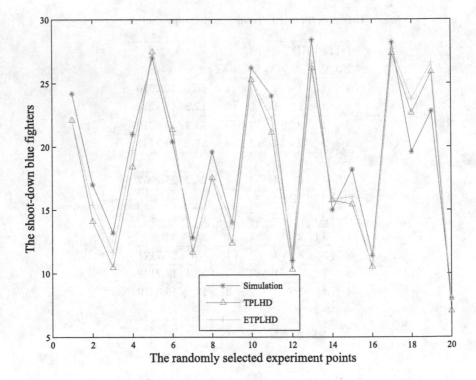

Fig. 5. The simulation and two prediction values of random points

4   Conclusion

In order to break the limitation of the fixed number of design points in traditional LHD method, a new flexible DOE method ETPLHD is proposed in this paper. The comparison shows that ETPLHD method performs better than TPLHD for no more than 6 variables. The ETPLHD method is also applied in a complex combat simulation to help to find proper approximation model of the combat system, which is used to predict the outcomes of the combat with different parameter configurations of the SAM missiles. The results show that the approximation model using the ETPLHD method has better predicting accuracy than the model using the TPLHD method.

References

1. Sanchez, S.M., Wan, H.: Work smarter, not harder: a tutorial on designing and conducting simulation experiments. In: Proceedings of the 2012 Winter Simulation Conference, pp. 1–15 (2012)
2. Ghosh, S.P., Tuel, W.G.: A design of an experiment to model data base system performance. IEEE Trans. Softw. Eng. **SE-2**(2), 97–106 (1976)
3. McKay, M.D., Beckman, R.J.: A comparison of three methods for selecting values of input variables from a computer code. Technometrics **21**(2), 239–245 (1979)
4. Kelton, W.D.: Designing simulation experiments. In: Proceedings of the 1999 Winter Conference, pp. 33–38 (1999)
5. Park, J.S.: Optimal Latin-hypercube designs for computer experiments. J. Stat. Plann. Inference **30**(1), 95–111 (1994)
6. Morris, M.D., Mitchell, T.J.: Exploratory designs for computational experiments. J. Stat. Plann. Inference **43**(3), 381–402 (1995)
7. Bates, S.J., Sienz, J., Toropov, V.V.: Formulation of the optimal Latin hypercube design of experiments using a permutation genetic algorithm. In: 45th AIAA/ASME/ASCE/AHS/ASC Structures, Structural Dynamics and Materials Conference, AIAA-2004-2011, Palm Springs, California, pp. 19–22 (2004)
8. Viana, F.A.C., Venter, G., Balabanov, V.: An algorithm for fast optimal latin hypercube design of experiments. Int. J. Numerical Methods Eng. **82**(2), 135–156 (2009)
9. Ye, K.Q., Li, W., Sudjianto, A.: Algorithmic construction of optimal symmetric latin hypercube designs. J. Stat. Plann. Inference **90**, 145–159 (2000)
10. Ma, Y., Gong, G.: A research on CGF reasoning system based on Fuzzy Petri Net. In: Proceedings of System Simulation and Scientific Computing, Beijing, pp. 406–411 (2005)

Modeling Behavior of Mobile Application
Using Discrete Event System Formalism

Yun Jong Kim, Ji Yong Yang, Young Min Kim, Joongsup Lee,
and Changbeom Choi[✉]

Handong Global University, Newton 206 558 Handong-ro Buk-gu, Pohang,
Gyeongbuk 37554, Republic of Korea
{21100132,21000416,21100114,joongsup,cbchoi}@handong.edu

Abstract. Discrete Event System Specification (DEVS) is a set theoretic formalism developed for specifying discrete event systems. DEVS features the specifications of hierarchical system and sequential event. Mobile application system is one of the typical systems that have hierarchical structure and sequential processes. However, there has been no attempt to express the mobile application system as DEVS formalism. This paper describes the design and development of mobile application behavior using DEVS model and we have classified the types of the mobile app behaviors as: temporal behavior, non-temporal behavior. Each behavior is stated as an atomic model and the mobile app events are represented as a coupled model. By using the DEVS formalism, not only we can simulate most of the mobile application events using real device; but also we expect that this work will serve as an effective tool for usability test.

Keywords: DEVS · Mobile application · Simulation environment · DEVS modeling · Model implementation · Real device

1 Introduction

According to a new market research report published by 'VisonMobile', the total global mobile applications market is expected to be worth US$58.0 billion by the end of 2016. The market of the mobile application is growing fast and user's requirement is also becoming diversified. To further, there have been many studies and attempts to find solutions that reflects the user's needs efficiently.

One of the biggest challenges in mobile application development is that the requirements for mobile application change frequently and rapidly [1]. Many developers have difficulties to keep the code up to date and it takes a lot of time. To deal with this problem, rapid prototyping has been used [2]. In this methodology, research and development stages are conducted in parallel to another when creating the prototype [3]. This enables developer and designer to acquire precise feedback and analyze user requirements by providing high-fidelity prototypes.

In order to develop a high-fidelity prototype, the developer should consider the behavior of mobile application. The typical behavior of mobile application can be in turn categorized into temporal behavior and non-temporal behavior. The temporal

© Springer Science+Business Media Singapore 2016
S.Y. Ohn and S.D. Chi (Eds.): AsiaSim 2015, CCIS 603, pp. 40–48, 2016.
DOI: 10.1007/978-981-10-2158-9_4

behavior does not require users' input, but instead it is triggered by external sources. Display of animation on mobile screen and pop-up message from mobile applications are such examples. On the other hand, non-temporal behavior is the event that requires user's input. Screen touch, swipe and long touch are some examples of this behavior. We have applied discrete event system specification formalism to monitor these behaviors.

This paper contains five sections. Section 2 describes the related works concerning the making of high-fidelity mobile application prototype. Section 3 introduces how to model the temporal behavior and non-temporal behavior using DEVS formalism, which sets the theoretical background of our study. Furthermore, we explain the overall structure of the simulation using DEVS formalism. Section 4 introduces the case study with the proposed approach and we conclude our study in Sect. 5.

2   Related Works

Mobile application can be grouped by the three categories: native, web-based, and hybrid [4, 5]. All three types of mobile applications needs prototyping process under the development in common. To support the process, many prototyping tools currently exist in the market.

The table below shows the limits of existing tools. We compared four well-known prototyping tools with our proposed approach. There are several prototyping tools that support making of high-fidelity prototype. These tools simulate the mobile application by using the expected behavior information embedded in the pre-designed image. Note the existing tools only support the non-temporal behavior. Also, it is not possible to discrete each event with formal representation for the user to obtain additional information about their prototype (Table 1).

Table 1. The comparison between existing prototyping tools and proposed system

	Marvel	POP	Fluid	Axure	Proposed system
Modeling non-temporal behavior	O	O	O	O	O
Modeling temporal behavior	X	X	X	X	O
Formal representation	X	X	X	X	O

3   Modeling and Simulation of Mobile Application Formalism

3.1   Discrete Event System Specification Formalism

The DEVS formalism supports the structure of a system in modular and hierarchical point of view. There are two kinds of DEVS model: one is atomic model and the other is coupled model [6]. While a part of system can be represented as an atomic model, a system represented in DEVS coupled model shows how a system can be coupled together and also how they interact with each other.

The coupled model is composed of various models. This may express larger system by taking an atomic model or a subordinated coupled model as a child model. A coupled model 'CM' is defined as:

$$CM = < X, Y, \{M_i\}, EIC, EOC, IC, SELECT>$$

X	set of internal input events;
Y	set of external output events;
$\{M_i\}$	set of components
EIC	external input coupling
EOC	external output coupling
IC	internal coupling
SELECT	the tie-breaking function to arbitrate occurrences of simultaneous events

The Atomic model is the most basic module in the hierarchical structure. 'M_i' in the CM corresponds to the atomic model. It describes the system behavior. An atomic model 'M' is defined as follows:

$$M = < X, S, Y, S, \delta_{ext}, \delta_{int}, \lambda, ta > \text{ where}$$

X	set of internal input events
S	set of state
Y	set of external output events
δ_{ext}	external transition function specifying state transitions based on the external events
λ	output function generating external events as output
ta	time advance function

Mobile application can be viewed as the collection of various components and this can be expressed by using DEVS formalism models. To explain in detail, we use the term 'Activity' in Android, which is an application component that provides a screen for the users to interact. As shown in Fig. 1, one activity consists of many mobile

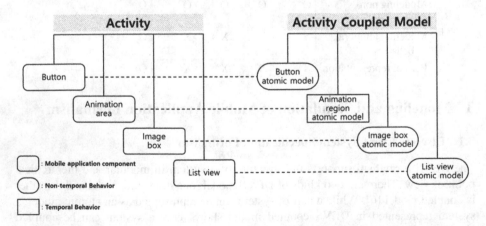

Fig. 1. The composition of mobile application activity with DEVS formalism model

application components. Each mobile application component corresponds to the proper atomic model concerning the type of behavior. For example, all the behavior related to the button component is the non-temporal behavior model and this can be expressed as an atomic model describing non-temporal behavior. On the other hand, the behavior in which the animation is displayed on screen can be represented as an atomic model and classified as temporal behavior. Such activities have various components that correspond to the coupled model of many atomic models.

3.2 Modeling Non-temporal Behavior of Mobile Application

Figure 2 shows how the non-temporal behavior model works. In this model, there are three states: QUEUE, WAIT and SEND. The initial state is WAIT and this takes place when the system is ready to receive the input event. As the non-temporal behavior model receives the event data, the system turns into the SEND state. The model sends out an execution message through $Output_1$ through state transitions from SEND to QUEUE or WAIT. When the queue is empty, the system becomes WAIT; and when the queue size is above zero, it reaches the QUEUE state. Overall, this model generates non-temporal event, which in turn can be performed in the Execution model, which we will cover in Sect. 4.

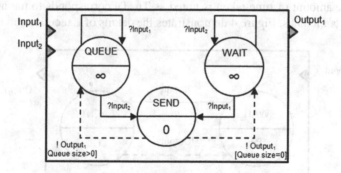

Fig. 2. The details of non-temporal behavior model

3.3 Modeling Temporal Behavior of Mobile Application

Figure 3 describes how the temporal behavior model works. This study conveys one of temporal behavior examples: the animation model. The animation model simulates the gradual change of specific area of the screen without user input. This model has three states: WAIT, PROCESS and RESULT. At the initial state WAIT, the animation area displays the initial image. Internal transition function occurs after T_W seconds and the state of the model changes into PROCESS. For each stay, it spends T_P seconds. Temporal behavior model stays at PROCESS state N times by increasing count. This means the model invokes the output function N times and gradually changes the image for each function-call. When the count becomes equal to N, the system reaches the RESULT state. In this stage, the model changes the screen image into the last one that is to be shown in the animation. After T_R seconds, the state goes back to the initial state and the cycle is repeated.

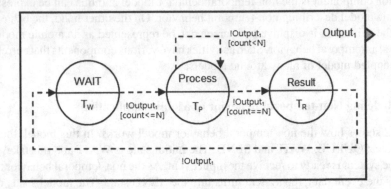

Fig. 3. The details of temporal behavior model

3.4 Modeling Execution Model

The execution model sends output message to implement specific event after given time, Tn. Note the amount of time taken is noted as Tn. Tn corresponds to the number that the event data includes. Figure 4 demonstrates the details of Execution model.

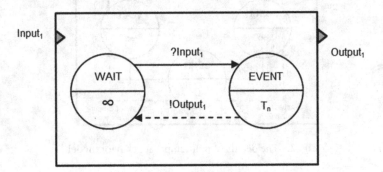

Fig. 4. The details of execution model

There are two states: WAIT, EVENT. The initial state is WAIT. Once the message is arrived from the behavior atomic model, the state changes to EVENT. Consequently, the model turns itself back into WAIT state through the internal transition and sends message allowing the mobile OS to operate.

3.5   The Mobile Simulation Using DEVS Formalism

The simulation architecture using DEVS formalism that this paper suggests is depicted in Fig. 5. This is composed of three parts; Mobile OS, Event convertor for mobile application (ECMA) and DEVS model describing the behavior of mobile application. This system receives the input event that is going to be simulated through the mobile OS. Basically, the Input event means the event time and the type of event. These events are transferred to DEVS formalism models through ECMA. The input message is transferred through callback function in the 'input event convertor'. When the behavior atomic model receives the input message, the system generates the output message and the execution model is transferred into DEVS message. The output event convertor in ECMA then transfers the message via mobile OS by making system API call for each type of event.

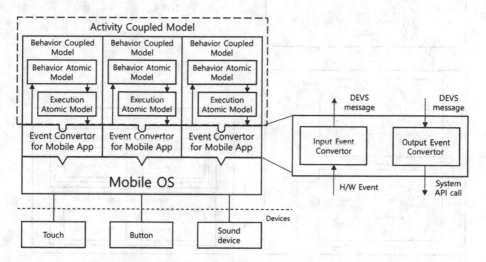

Fig. 5.   The simulation structure of mobile application

4   Case Study

In this section, we introduce two specific events which can be expressed by using the behavior models we have proposed earlier. Note various features of simulation using the behavior model can be seen from the case study. The events that we simulate are twofold: (1) Activity conversion by button touch and (2) Image conversion with the lapse of time.

4.1   Activity Conversion by Button Touch

Activity conversion by touch button is one of the events that is triggered by the user's input. Figure 6 describes the event with DEVS formalism models. This event can be expressed as the coupled model, which comprises the non-temporal behavior model and the execution model. The execution model of non-temporal behavior enables the conversion to occur and as the result, the image displayed on screen changes into the upper-left image shown in Fig. 6.

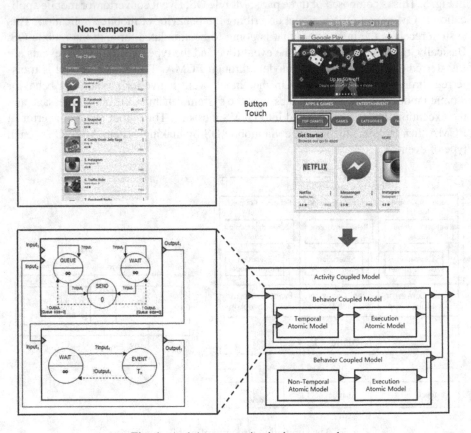

Fig. 6. Activity conversion by button touch

4.2 Image Conversion with the Lapse of Time

The image conversion event in Fig. 7 requires current activity as an input source. Because the direct user input is unnecessary for this event, the event in Fig. 7 can be expressed as the coupled model, which comprises the temporal behavior model and the execution model. The execution model takes some time as temporal behavior model, and then changes the screen image into the upper-right image shown in Fig. 7.

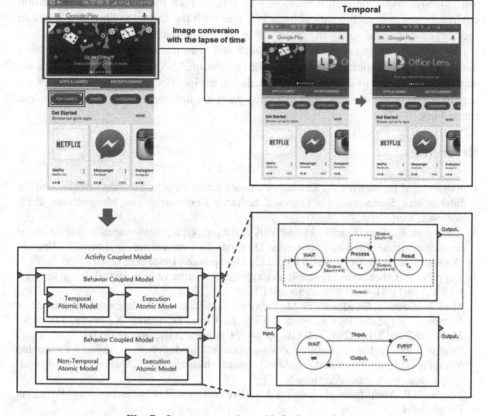

Fig. 7. Image conversion with the lapse of time

5   Conclusion

Mobile application system is a highly hierarchical and time-based process [7]. Because of these characteristics, DEVS formalism is one of the appropriate ways to model mobile application system. However, there has been no attempt to express mobile application workings through DEVS.

This study simulates the mobile application by stating the mobile app behavior in DEVS models and introduces how to model the mobile app behavior. We classified by the type of the mobile app behavior; and made models of such behavior using the DEVS atomic model. By combining the behavior atomic model and coupled model we can state the various mobile application events.

The approach that we proposed represents the temporal behavior which is not possible to state in existing prototyping tools. This helps to implement the prototype with high fidelity. In addition, the mobile app simulation using DEVS model enables to accumulate formal data. This means that the simulation environment that we are proposing can in turn be used to test usability by analyzing the accumulated data. Many other industries utilize the data from the prototype to analyze the task-based usability test [8]. Moreover, the formal method, which is ensured by DEVS based simulation environment, has a mathematical basis [9]; thus with the formal method, the user can specify, develop, and verify the prototype in systematic manner.

However, the implementation framework still does not exist; hence the user must develop the entire simulator model from the bottom up. Also, simulator heavily depends on platform because of the absence of simulation engine. For further research topics we may look into simulation engine and finding ways to implement the device-independent framework.

References

1. Joorabchi, M.E., Mesbah, A., Kruchten, P.: Real Challenges in Mobile App Development. In: International Symposium on Empirical Software Engineering and Measurement, IEEE computer society, pp. 15–24 (2013)
2. Adelmann, R., Langheinrich, M.: SPARK rapid prototyping environment – mobile phone development made easy. In: Tavangarian, D., Kirste, T., Timmermann, D., Lucke, U., Versick, D. (eds.) IMC 2009. CCIS, vol. 53, pp. 225–237. Springer, Heidelberg (2009)
3. Tripp, S.D., Bichelmeyer, B.: Rapid prototyping: an alternative instructional design strategy. ETR&D **38**(1), 31–44 (1990). ISSN 1042-1629
4. Masi, E., Cantone, G., Mastrofini, M., Calavaro, G., Subiaco, P.: Mobile apps development: a framework for technology decision making. In: Uhler, D., Mehta, K., Wong, J.L. (eds.) MobiCASE 2012. LNICST, vol. 110, pp. 64–79. Springer, Heidelberg (2013)
5. "Native, web or hybrid mobile-app development," IBM Software, Thought Leadership WhitePaper. http://www.computerworld.com.au/whitepaper/371126/native-web-or-hybrid-mobile-app-development/download/
6. Zeigler, B.P.: Multi-facetted Modeling And Discrete Event Simulation. Academic Press, San Diego (1984)
7. Zeigler, B.P., Chow, A.C.H.: Parallel DEVS, hierarchical, modular, modeling formalism (1994)
8. Stone, B., Wang, Y.: AirportLogic: usability testing, prototyping, and analysis of an airport wayfinding application. In: Stephanidis, C. (ed.) Posters, Part II, HCII 2011. CCIS, vol. 174, pp. 81–84. Springer, Heidelberg (2011)
9. Seo, K.-M., Choi, C., Kim, T.G., Kim, J.H.: DEVS-based combat modeling for engagement-level simulation. Simul. Trans. Soc. Model. Simul. Int. **90**(7), 759–781 (2014)

Multi-modal Transit Station Planning Method Using Discrete Event System Formalism

Jaewoong Choi, Bitnal Kim, Onyu Kang, Seonwha Baek,
Yonghyun Shim, and Changbeom Choi[✉]

Handong Global University, Newton 206 558 Handong-ro Buk-gu,
Pohang, Gyeongbuk 37554, Republic of Korea
{21000749,21100089,21400008,21400358,21100384,
cbchoi}@handong.edu
http://www.handong.edu

Abstract. A multi-modal transit station is a complex system in the urban system and takes important roles in our daily life. Since various transportation systems a large scale of passenger participates to the station, design and analysis of the multi-modal transit station are important problem. This paper presents a modeling and simulation method to support designing the transit station. First, we introduce the modeling method to build a simulation models to capture the behavior of vehicles and the crowds using the Discrete Event System (DEVS) formalism. Then, we introduce a simulation environment for multi-modal transit station which is based on the DEVSim++. In this simulation environment, we support requirement verification method based on the untimed DEVS model, and it checks the requirements by executing the simulation model and the untimed DEVS model altogether.

Keywords: Transportation simulation · DEVS formalism · Agent based simulation

1 Introduction

A transportation takes an important role to support a member of modern society to move places to places. Especially, rapid urbanization has been changed a citizen's daily life dramatically. In urbanized society, large scale of the citizen utilizes various transportation method to move from home to their work places, and vice versa. As the urbanization accelerates, the importance of designing the multi-modal transit station increases. In multi-modal transit station, various transportation arrives to the station with its own time schedule and a large scale of passenger may arrive to the station to utilize a transportation. As a result, an urban planner or an architect should consider several consideration points, such as traffic, volume of the passenger flow, and other transit modes to design the multi-modal transit station. If an urban planner underestimates the traffic flow to the multi-modal transit station, the access roads to the multi-modal transit station will be too narrow to handle the traffics. On the other hand, when the planner overestimates the traffic flow, the size of the access roads will be too wide, and it will be waste of the

© Springer Science+Business Media Singapore 2016
S.Y. Ohn and S.D. Chi (Eds.): AsiaSim 2015, CCIS 603, pp. 49–64, 2016.
DOI: 10.1007/978-981-10-2158-9_5

resources. Moreover, it should spend a huge budget and time to fix such problems after the station has been built. Modeling and simulation is one of the alternatives to help the planner to design multi-modal transit station before the construction. By using modeling and simulation, an urban planner may consider the traffic flow and volume of the passenger flow before building the multi-modal transit station. Also, the planner may consider the geographical layout of access roads to the transit station and the behavior of the individual passenger to predict the traffic flow and volume of the passenger flow.

Modeling and simulation of urban planning is nothing new to urban planning domain. There were several studies that utilize the modeling traffic and the layout of the road in the cities and simulate them to acquire data to find the congestion point. Especially, a traffic simulation model is widely used to analyze the alternatives of transportation system without directly changing the real world, such as constructing additional road or modulating traffic signals [1, 2]. In order to capture such behaviors of traffics, passengers, and pedestrian, two modeling and simulation methods have been proposed: macro simulation model and micro simulation model. Both models were used to capture the characteristics of the traffic of a transportation system. However, the previous studies were used to predicting the traffic situations. Also, to the best of the authors' knowledge, there are no studies that confirm whether the design of multi-modal transit station satisfies the requirements or not.

In this paper, we propose Discrete Event System (DEVS) Formalism based modeling method to capture the behavior of the vehicle and passenger. Also, we propose the simulation environment to check the transit station model against to the requirement. The DEVS Formalism is set theoretical formalism which has hierarchical structure and modular feature. First, we use model of the DEVS Formalism, the DEVS model, to model spatial layout of the multi-modal transit station in order to capture the behavior of the vehicles and passengers' movements. Especially, we model the spatial layout by connecting DEVS model consecutively. Also, we model temporal and non-temporal behaviors of the vehicle and passenger as discrete events from previous DEVS model to next DEVS model. Also, we model the interaction between vehicle and passenger by exchanging discrete event from DEVS model to DEVS model.

Second, we propose simulation environment that supports various requirement specification methods to check the model against to its requirements. In this paper, we mainly utilize requirement specification method based on the propositional logic. However, in order to support various requirement specification methods, we unify the modeling method and the requirement specification method using DEVS model. The model for requirement specification continuously monitors the event sequences from model of the multi-modal transit station, and checks the temporal and non-temporal requirements. A preliminary version of this paper appeared in AsiaSim '15. In the previous paper, there are no details in modeling method and simulation environment. This paper is an extended and improved version of the preliminary version. The rest of this paper is organized as follows. Section 2 presents related works. In Sect. 3 introduces a modeling method for multi-modal transit station, and Sect. 4 presents the simulation environment for multi-modal transit station. Finally, Sect. 5 shows the case study of our proposed method and we conclude this paper.

2 Related Works and Background

In this section, we introduce the previous researches in modeling and simulation in urban planning domain and modeling and simulation theories. The previous researches can be categorized by macro simulation and micro simulation. The macro simulation model captures the characteristics of transportation system in macroscopic point of view rather than modeling individual elements. On the other hand, the micro simulation model models each individual transportation element to analyze the interaction among various transportation elements.

2.1 Related Works

The macro simulation model abstracts the type of the vehicles into the differential equations and captures the macroscopic characteristics, such as aggregated traffic flow dynamics and density of the traffic [3, 4].

A macro simulator has advantage on understanding traffic flows, but hard to predict behavior of each object. For example, a macro simulation model DYNEMO was designed for the development, evaluation and optimization of traffic control systems for motorway networks [4]. Traffic flow model included with the simulation package combines the advantages of a macroscopic model (computational simplicity) with the advantages of a microscopic model (output statistics relating to individual vehicles).

On the other hand, most studies about micro simulations were concentrated on individual vehicles, crossroads or traffic signal based simulations. For example, Errampalli et al. has provided a basis for evaluating the public transport policies such as public transport priority systems by developing microscopic traffic simulation model [2]. They have considered two types of public transport policies for evaluation. One is bus lanes and another is public transport priority systems at traffic signal. Fuzzy logic reasoning has been incorporated in route choice analysis while choosing the route based on the level of satisfaction of the available routes. The developed simulation model has been applied on the part of Gifu city network and it is able to predict the vehicular movements with a fair amount of accuracy. From this study, it can be concluded that the developed microscopic simulation model can be applied to evaluate public transport polices particularly bus lane and public transport priority systems at intersections with fair amount of accuracy.

Among various modeling and simulation theories in micro simulation, cellular automata and agent based simulation are the representative theory that model the transportation system and simulate them [5]. Especially, Zamith et al. has introduced a flexible and robust traffic cellular automata model [5]. They consider the motion expectation of vehicles that are in front of each driver. Then, they define how a specific vehicle decides to get around, considering the foreground traffic configuration. The model utilizes stochastic rules for both situations, using the probability density function of the beta distribution to model three drivers behavior, adjusting different parameters of the beta distribution for each one. The three drivers are as follows. First is Hunter and Tailgater that describes the behavior of a driver that intends to stay close to the upfront vehicle and wants to go fast. Second is Ultraconservative that describes the driver with the opposite profile of the Hunter and Tailgater. This driver prefers to keep a great distance to the upfront vehicle and drives slower. Third is Flow Conformist that describes a driver that intends to go according with the flow and is not characterized by none of the previous profiles. It models only drivers' behavior and it means they focus on cars.

Above mentioned previous studies have solid theoretical background to model the transportation network or the transportation systems. Our approach is similar to the microscopic simulation. Especially, we utilize the cellular automata to model the spatial layout and agents to denote the elements of the transportation system. However, our approach is differentiated by utilizing cellular automata and agents to model the multi-modal transit station. In our approach we focus to the spatial layout of the multi-transit station and we model the vehicles and the passengers as agents. Also, we model the boarding in the transit station by modeling the interaction among the vehicles and the passengers.

2.2 Discrete Event System Formalism

The DEVS formalism is a set-theoretic formalism developed for specifying discrete event systems [7–9]. In the DEVS formalism, one can specify basic models and how these models are coupled in a hierarchical and modular fashion. A basic model, called an atomic model, specifies the dynamics of a model and is defined as Atomic Model and Coupled Model.

An Atomic Model is defined as follow:

$$AM = <X, Y, S, \delta_{ext}, \delta_{int}, \lambda, ta>$$

where
   X: a set of external input event types,
   Y: an output set,
   S: a sequential state set,
   $\delta_{ext}: Q \times X \to S$, an external transition function
        where Q is the total state set of
        $M = \{(s, e) | s \in S \; and \; 0 \le e \le ta(s)\}$,
   $\delta_{int}: S \to S$, an internal transition function,
   $\lambda: S \to Y$, an output function,
   $ta: S \to \mathcal{R}_{0,\infty}^{+}$, a time advance function, where the $\mathcal{R}_{0,\infty}^{+}$
        is the non-negative real numbers with ∞ adjoined.

An atomic model AM is a model which is affected by external input events X and which in turn generates output events Y. The state set S represents the unique description of the model. The internal transition function δ_{int} and the external transition function δ_{ext} compute the next state of the model. If an external event arrives at the elapsed time e which is less than or equal to $ta(s)$ specified by the time advance function ta, a new state s' is computed by the external transition function δ_{ext}. Then, a new $ta(s')$ is computed, and the elapsed time e is set to zero. Otherwise, an internal event arrives at $ta(s)$, and then a new state s' is computed by the internal transition function δ_{int}. In the case of internal events, the output specified by the output function λ is produced based on the state s, which means the output function is processed before the internal transition function. Then, as before, a new $ta(s')$ is computed, and the elapsed time e is set to zero.

A Coupled Model is defined as follow:

$$CM = <D, \{M_i\}, \{I_i\}, \{Z_{i,j}\}, SELECT>$$

where

D: a set of component names,

For each i in D,

$\quad M_i$: a component basic model

$\quad$ (an atomic or coupled model),

$\quad I_i$: a set of influences of i,

and for each j in I_i,

$\quad Z_{i,j}: Y_i \rightarrow X_j$, an i-to-j output translation,

$SELECT: 2^M - \phi \rightarrow M$, a tie-breaking selector.

A coupled model CM consists of components $\{M_i\}$, which are atomic models and/or coupled models. The influences $\{I_i\}$ and i-to-j output translations $\{Z_{i,j}\}$ define the coupling specification as follows: An external input coupling (EIC) connects the input event of the coupled model to the input event of one of its components; An external output coupling (EOC) connects the output event of a component to the output event of the coupled model; An internal coupling (IC) connects the output event of a component to the input event of another component. The $SELECT$ function is used to order the processing of simultaneous internal events for sequential simulation. Thus, all the events with the same time in a system can be ordered by this function.

Also, the DEVS formalism has various implementations that supports the discrete event simulation. One of the popular implementation for DEVS formalism is the DEVSim++, which is implemented base on C++ programming language. The DEVSim++ coordinates the event schedules of atomic models in a system and provides classes and application programming interface for simulation. Since the DEVSim++ fully utilizes the C++ features, it provides the advantages of object-oriented framework, such as encapsulation, inheritance, and re-usability and it is utilized to implement various simulators in various domains [10–12].

3   Modeling Method for Multi-modal Transit Station

In this section we introduce modeling method for multi-modal transit station. A multi-modal transit station is complex system consist of different types of area. For example, a modeler should consider the transportation region and passenger region. Following Fig. 5 shows compositions of general multi-modal transit station. As shown in the Fig. 5, when a group of vehicles arrives to the station it is categorized by transportation type and it flows in to the waiting lane. After that, it waits on the waiting lane until the vehicle picks up a passenger, then it leaves the transit station through the exit roads. On the other hand, when a passenger arrives to the multi-modal transit station, a passenger decides the transportation method and move to the right platform. Therefore, a passenger uses the crosswalk to change the platform and a vehicle should wait for the passenger to cross the road (Fig. 1).

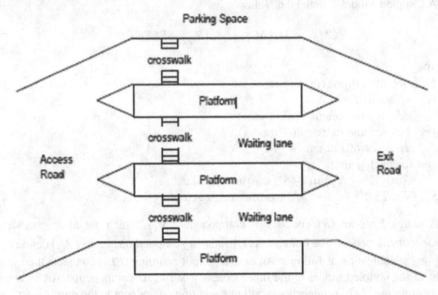

Fig. 1. Representation of general multi-model transit station

In our approach, we model spatial layout of the multi-modal transit station as an array using a formal representation. Then we model passenger and vehicle as an agent interacting inside of the multi-modal transit station model. Then, we model the procedure of taking taxi as interaction between passenger and vehicle models. In order to model each cell, we utilized the DEVS formalism.

3.1 Modeling Spatial Layout of the Station

A spatial layout of multi-modal transit station is one of the reason that causes traffic congestion, so it is one of the important consideration for designing multi-modal transit station. In order to model a spatial layout of the station, we divide a region into cells and each cell represents the characteristics of the region. For instance, a single lane which can hold 10 vehicles can be modeled as 10 cells. Then, each cell represents one vehicle and a modeler can model the behavior of the single lane by array of cells that models the characteristics of the single lane. Modeling a spatial layout of the multi-modal transit station using cells have two merits. First, a modeler can decide the resolution of the simulation by choosing the number of cells. When a modeler uses massive cells to model the transit station, the model can simulate the movements of passengers and vehicles in micro scale with high computation power. On the other hand, when a modeler wants to see the result quickly with the proper resolution, the modeler can reduce the number of cells. Therefore, a modeler first select a region in the multi-modal transit station, and selects the proper simulation resolution with respect to the simulation objective. Finally, each sub-region is mapped to a cell to model the region. Following Fig. 2 shows the representation of a road using cells. As shown in the Fig. 2, if a modeler wants to model a road, the modeler should decide the resolution of the simulation. In this case, five cells represent a sub-region of the given road with 10 m.

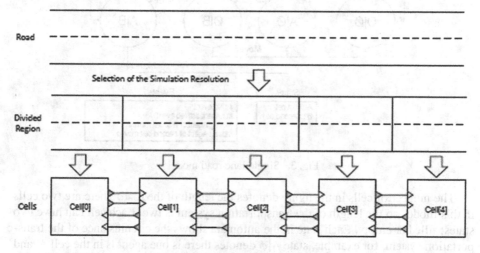

Fig. 2. Cell representation of a road

To model the region by dividing smaller region to model the behaviors of the sub-region, we adopted the cellular automata theory [13]. Therefore, we can model the road for a vehicle and a passenger by cells that interact with each other.

3.2 Modeling Behavior of Agent

In the real world, a passenger and a vehicle in the multi-modal transit station show certain tendency to move from one place to other place. They are under controlled by rules which are given from the society. For example, if we want to model a multi-modal transit station with taxi, a passenger and a vehicle should follow following rules: "A taxi cannot overtake the preceding taxi in the single lane." and "When there is a passenger waiting in front line, a passenger from backside of the line should not take a taxi". Such rules are the propositional statements that the simulation models of the multi-modal transit station must be followed. A proposition is a declarative sentence that is either true or false, but not both [14]. Also, we can develop a proposition by connecting various propositions using operators, such as conjunction, disjunction, and negation. As a result, we can develop a complex proposition by combining propositions to give requirement specification to a given system. Such propositions should always be true during the simulations. As a result, an urban planner can use propositional logic to check the violation of the requirement specifications by monitoring the events during the simulation, since some events sequence may lead the system into violation status, i.e., some circumstance that the given propositions are not true.

Also, the behaviors of the vehicle and the passenger can be captured by automata-theoretic approach with respect to the propositional logic [15]. Following Fig. 3 shows the simple single lane road system.

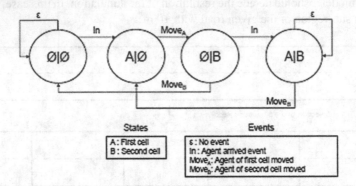

Fig. 3.   Single lane road model

The number of cells in the figure denotes the length of the road. There are two cells in this model, so the length of the transportation system is two. Each cell can have two states: filled or empty. Each state in the automata shows the circumstance of the transportation system, for example, state $A|\phi$ denotes there is one agent is in the cell A, and $A|B$ denotes there are two agents in the system. Also, four events are defined: ϵ, In, $Move_A$, and $Move_B$. Each event denotes, no events are occurred, an agent has arrived to the system, an agent have departed from cell A, and an agent have departed from cell B, respectively.

Since we can consider the passenger line as a queue, we can use automata to represent the behavior of the passengers. Also, since there are two lanes, one for the taxis and other for the passengers, we should cross product two automata to represent the combined behavior between taxis and passengers. Figure 4 shows the cross product of

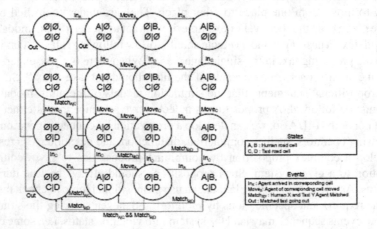

Fig. 4.   Cross product of two single lane transportation system

the two single lane transportation system. As shown in the figure, all transitions are the valid behaviors in the multi-modal transit station. Therefore, a user can detect the violation of the requirement specifications by monitoring the transitions of the transportation system.

3.3   Modeling Interaction Between Vehicle and Human

To model the multi-modal transit station, an urban planner should consider the vehicle road and the human road. Each road may contain vehicle agents and passenger agents, respectively. Since we are modeling the multi-modal transit station, each passenger interacts with vehicle. When a vehicle can ride a passenger, a passenger will disappear from the human road.

Figure 5 shows a boarding in the multi-modal transit station. In the multi-modal transit station, boarding process shown in Fig. 5 is occurred multiple times in parallel. Since both of the roads are the queuing model, simulation model of the vehicle and the passenger can be considered agents that wait in the line. The behavior of each roads is similar, each agent in the road should follow the lane and it cannot overtake the agent in the front. In addition, we can model the boarding process by matching proper pair of the vehicle and the passenger. Figure 6 shows the valid transitions of the automata for a multi-modal transit station considering the matching between a vehicle and a passenger.

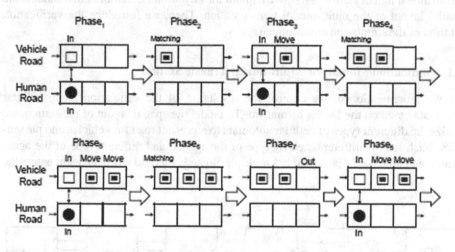

Fig. 5. Boarding process in a multi-modal transit station

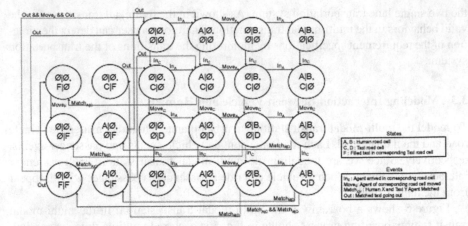

Fig. 6. Automata of the multi-modal transit station considering matching

4 Simulation Environment of the Multi-modal Transit Station

Building a simulation model for multi-modal transit station is not an easy task, especially to the urban planners. In this section we introduce the simulation environment of the multi-modal transit station. First, we explain the key simulation model that models the spatial layout of the multi-modal transit station. Then, we introduce the verification method of the simulation environment.

4.1 Simulation Models for Multi-modal Transit Station

As we aforementioned, we control the resolution of the simulation using cellular automata based on the DEVS formalism. To model the spatial layout of the station, we utilize the different types of cellular automata to represent road for vehicles and passengers. Each road is differentiated by type of the agents, and different type of the agent cannot enter the road. Each road has multiple aligned cells and when an agent enters the

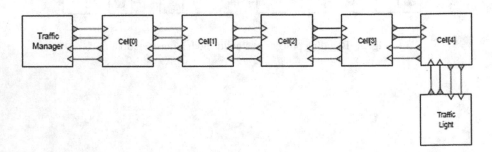

Fig. 7. Layout of the road models in DEVS formalism

road, it moves forward in their lane. In order to capture such semantics, we adopted one-dimensional cellular automata using DEVS formalism. Figure 7 shows the layout of each road model.

Each agent in the road model checks the state of the next cell and decides whether to move or stop. After the cells agent moves, the state of the cell is changed. This information is delivered to previous cell, and previous cell decides whether to go or stop. Therefore, the information propagates backward. The DEVS formalism is suitable to adopt this backward propagation mechanism.

The Fig. 8 shows the Coupled Model and Atomic Model which each cell checks the state of the next cell and decides whether to send agent to next cell or not. The difference between vehicle road model and human road model is the agent which enters to the model.

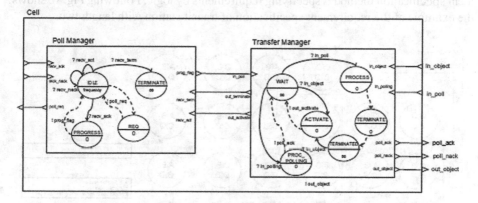

Fig. 8. Cell DEVS model in the transportation simulator

Figure 9 shows the Traffic Manager Model of the road model. The Traffic Manager Model generates the agents to the road, and urban planner may control the traffic flow by changing parameters in the model. The Traffic Manager Model has two model the Poll Manager model and Agent Generator model.

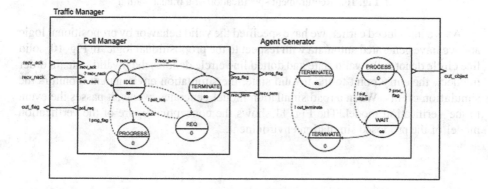

Fig. 9. Traffic manager model in the road model

The Agent Generator Model generates agent objects and sends it to first road cell. Each road cell has Poll Manager Model and Transfer Manager Model as shown in Fig. 8. When a Poll Manager Model sends state information of the cell to corresponding Transfer Manager Model when the state is changed. If a Poll manager gets this acknowledgement message, the road cell delivers agent object to next cell or matches the taxi and human. On the other hand, it will be blocked because other agent is existing in the front cell.

4.2   Verification Model for Multi-modal Transit Station

To check the simulation model that satisfy the requirement specification, an urban planner must provide rigorous requirement specification. One of the rigorous requirement specification method is specifying requirements by logic. Following Fig. 10 shows the example of the requirement specification of transit station with length two.

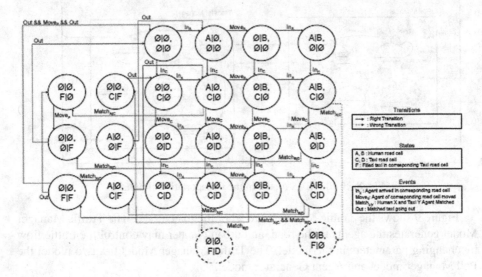

Fig. 10. Requirement specification of a transit station

As we introduced earlier, we have specified the valid behavior by propositional logic and we have generated automata with respect to the propositional logic. In Fig. 10, solid line circle denotes the valid state and dotted line circle denotes the invalid state. In order to check the entry point to the invalid state, the simulation environment monitors the simulation events. When a road simulation model generates events, it passes the event to the verification model. The Fig. 11 shows the conceptual figure of the verification model of the proposed simulation environment.

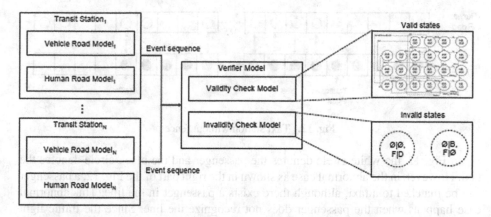

Fig. 11. Verification model of the proposed simulation environment

The verification model has two models, the Validity Check Model and Invalidity Check Model. The Validity Check Model continuously monitors the event sequences from the transit station model to check the transition model that generates valid events. On the other hand, the Invalidity Check Model only check the invalid events that cannot be existed through the simulation. For example, the Invalidity Check Model check the events that lead to the abnormal states, which is matching has been occurred at the back of the line despite empty vehicle exist at the front or customer has ridden to the vehicle despite the customer at the front did not ride a vehicle.

5   Case Study: Taxi Station of Pohang KTX Station in South Korea

This section introduce a case study of taxi station of Pohang KTX (Korean Train eXpress) Station in South Korea. The Pohang KTX station has opened in 2015, and many of the passenger had suffered terrible traffic jam while using transit station. One of the reason for traffic jam was the Taxi station, so we have built an abstracted simulation model to model the taxi station of Pohang KTX station. The main objective of the simulator is to check the fairness of the taxi station system. In order to check the fairness of the system, we first specify the requirement specifications for the taxi and the passengers. Following propositions are the requirements of the tax station.

Proposition 1. *A taxi cannot overtake the preceding taxi in the single lane.*

Proposition 2. *A passenger must be served in a First Come First Serve (FCFS) fashion.*

Based on the propositions, a passenger should be served as FCFS fashion and the Taxi should leave the transit station one by one. By simulating the model, we have found the following situations: (1) ideal case and (2) abnormal case. The ideal case, every passenger and the taxi are matched in FCFS fashion, as shown in the left part of the Fig. 12.

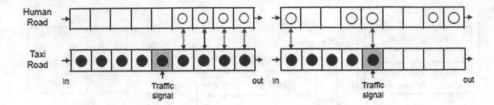

Fig. 12. Taxi station without fence

In Fig. 12, the white circle denotes the passenger and the black circle denotes the taxi. However, in the abnormal case as shown in the right part of the Fig. 12, a passenger can be matched to a taxi, although there exists a passenger in the line. This abnormal case happens when the passenger does not recognize the line. Since the traffic light prevents the taxi to enter the single lane, a passenger, who arrived to the taxi station, may think he/she is the first passenger in the line, so he/she will take a taxi although there exists a passenger in the line. We have also analyzed the alternative solution of the taxi station. Figure 13 shows the alternative solution, which is installing fence. In order to prevent random matching at the taxi station, a fence can be installed to the transit station. Fortunately, the decision maker of the Pohang KTX station has installed a fence to prevent a passenger to take a taxi anywhere. Figure 14 shows the initial phase of the taxi station and current taxi station of Pohang KTX station.

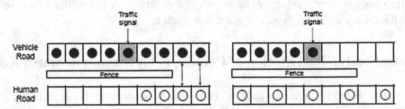

Fig. 13. Taxi station with fence system

(a) Taxi station at the initial phase   (b) Current taxi station

Fig. 14. Pictures of the taxi station of Pohang KTX station

6 Conclusion

Designing multi-modal transit station is one of important consideration in urban planning. Since urban infrastructure cannot be modified easily, a decision maker may consider the various alternative designs before building a transit station. To assist the urban planners, modeling and simulation of transit station is one of the most adequate method to draw alternatives efficiently.

In this paper, we presented a design method for multi-modal transit station and proposed a simulation environment based on the DEVS formalism. In order to capture the behavior of the vehicles and the passenger in the transit station, we model the spatial layout of the station as multiple cells using DEVS Formalism. Then, we implemented the simulation model using DEVSim++. Moreover, in order to verify the requirement specifications, we provided various method to check the requirement specification, such as propositional logic. To verify the requirement specification, we developed the untimed DEVS model to check the target transportation system against to the requirement specification. For the further research, a multi-lane system and temporal behavior of the traffic jam in the transportation network can be considered.

References

1. Hernndez, J., Ossowski, S , Garca-Serrano, A.: Multiagent architectures for intelligent traffic management systems. Transp. Res. Part C Emerg. Technol. **10**(56), 473–506 (2002)
2. Errampalli, M., Okushima, M., Akiyama, T.: Microsopic simulation model considering public transport policy. J. Eastern Asia Soc. Transp. Stud. **6**, 2718–2733 (2005)
3. Papageorgiou, M.: Concise encyclopedia of traffic & transportation system: simulation program, pp. 491–502 (1991)
4. Schwerdtfeger, T.: Dynemo: a model for the simulation of traffic flow in motorway networks. In: The 9th International Symposium on Transportation and Traffic Theory, pp. 65–87. VNU Science Press (1984)
5. Zamith, M., et al.: A new stochastic cellular automata model for traffic flow simulation with drivers' behavior prediction. J. Comput. Sci. **9**, 51–56 (2015)
6. OBrien, E.J., Lipari, A., Caprani, C.C.: Micro-simulation of single-lane traffic to identify critical loading conditions for long-span bridges. Eng. Struct. **94**, 137–148 (2015)
7. Zeigler, B.P.: Multifacetted Modeling and Discrete Event Simulation. Academic Press, London (1984)
8. Zeigler, B.P.: Object-Oriented Simulation with Hierarchical, Modular Models. Academic Press, London (1990)
9. Zeigler, B.P., Kim, T.G., Praehofer, H.: Theory of Modeling and Simulation. Academic Press, London (2000)
10. Song, H.S., Kim, T.G.: Application of real-time DEVS to analysis of safety critical embedded control systems: railroad-crossing control example. Simul. Trans. Soc. Model. Simul. Int. **81**(2), 119–136 (2005)
11. Hong, J., Seo, K.M., Kim, T.G.: Simulation-based optimization for design parameter exploration in hybrid system: a defense system example. Simul. Trans. Soc. Model. Simul. Int. **89**(3), 362–380 (2013)
12. Seo, K.M., et al.: DEVS-based combat modeling for engagement-level defense simulation. Simul. Trans. Soc. Model. Simul. Int. **90**(7), 759–781 (2014)

13. Wolfram, S.: Theory and Applications of Cellular Automata, vol. 1. World Scientific, Singapore (1986)
14. Rosen, K.: Discrete Mathematics and Its Applications, 7th edn. McGraw-Hill Education, New York (2012)
15. Vardi, M.: An automata-theoretic approach to linear temporal logic. In: Moller, F., Birtwistle, G. (eds.) Logics for Concurrency. LNCS, vol. 1043, pp. 238–266. Springer, Heidelberg (1996)
16. Tag Gon Kim, DEVSim++ Manual. http://smslab.kaist.ac.kr

Reliability Analysis of Software Defined Wireless Sensor Networks

Na Gong and Xin Huang[✉]

International Business School in Suzhou, Department of Computer Science
and Software Engineering, Xi'an Jiaotong-Liverpool University, Suzhou, China
Na.Gong12@student.xjtlu.edu.cn, Xin.Huang@xjtlu.edu.cn

Abstract. Software defined wireless sensor networks (SDWSN), which
separate the control and data forwarding functions, have been proposed
to solve the management dilemma of conventional wireless sensor networks (WSN). This technique can achieve a dynamic and centralized
network management. However, the study on its reliability is still defi-
cient. In this paper, SDWSN is formally modeled using Continuous-Time
Markov Chains (CTMCs), and its reliability is verified using probabilistic model checking techniques. The verification results are expected to
give suggestions to the future SDWSN designs.

Keywords: Model checking · PRISM · Software defined network · Wireless sensor network

1 Introduction

Nowadays, as an important technique in the next generation wireless and mobile
networks, the wireless sensor network (WSN) is becoming more and more complex and demanding [6,7,17]. It has been widely used in many areas, for example military, navigation and environmental monitoring [15]. Typical WSNs are
self-configuring networks [10,15], and their topologies are frequently changed
[11]. This feature of WSNs poses the demand for dynamic network management. Therefore, software-defined wireless sensor network (SDWSN), which is
a combination of WSN and software-defined networking (SDN), is proposed
[5,10,11,15,18,23].

SDN is a new paradigm of networking system that provides dynamic network management and configuration. Why SDN is needed? Traditional networks
consist of many devices for forwarding and processing data packets [21]. It is
extremely time-consuming to deploy and maintain traditional networks, because
the packet routing control functions are located in routers, and they need to be
updated locally in most cases [11]. To remedy this limitation, SDN is proposed.
It separates the control plane and the data plane. The data plane composed
by physical network devices is responsible to perform packet forwarding and
processing functions. The control layer is in charge of the whole network management [22]. The applications hosted in the application layer can program the
devices in data plane [14].

© Springer Science+Business Media Singapore 2016
S.Y. Ohn and S.D. Chi (Eds.): AsiaSim 2015, CCIS 603, pp. 65–78, 2016.
DOI: 10.1007/978-981-10-2158-9_6

Similar to SDN, SDWSN aims at separating the control plane from the data forwarding plane in WSNs. These days, the dramatically raising application of the wireless and sensing technology expands the requirement for the dynamic, flexible and secure management of WSN. Hence, to satisfy these needs, the Software-Defined Wireless Sensor Network (SDWSN) is proposed [5,6,11,15,18,23]. Compared with the traditional WSN, SDWSN has several advantages:

- The network becomes programmable. This allows for more dynamic network segmentation and utilization without changes in other devices [12].
- The centralized controller maintains a global view of the whole network. Thus, security and flexibility of SDWSN could be improved.

However, one major concern about SDWSN is its reliability. Because SDWSM has a central controller, it may becomes the bottleneck of the whole system. If it fails, the whole system will be down. However, so far, few researchers study the reliability of the SDWSN which however is not a negligible issue. This paper therefore aims to study the reliability of the SDWSN by using the probabilistic model checking technique with the tool PRISM.

The reminder of this paper is organized as follows. In Sect. 2, SDWSN is described. The probabilistic model checking technique is explained in Sect. 3. Section 4 describes the modeling of the SDWSN structure. Section 5 analyzes the results. Finally, Sect. 6 provides the conclusion.

2   Software Defined Wireless Sensor Network

The widespread application of WSN attracts much awareness from both the industrial and academic fields. Recently, many researchers focuses on enhancing the capability of WSN reconfiguration adaption. First of all, many MAC protocols such as T-MAC, C-MAC and snapMAC have been proposed [3,19,20]. Also, a centralized architecture is proposed for WSNs [9]. In addition, using SDN techniques in WSNs gains attentions of many researchers [5,6,10,11,15,18,23].

A SDN is composed by the following layers (shown in Fig. 1):

- Data plane: it consists of network devices such as switches to maintain the packet forwarding functionality [22].
- Controller plane: it is in charge of the whole network management especial the management of the packet processing rules [21].
- Application plane: it contains various applications such as Could Computing Service and Data Warehouse [1].
- APIs: it allows the centralized controller entity in the intermediate plane to transfer instructions and information from or to the data and application plane [4]. The OpenFlow protocol now is a popular one [12].

The construction of SDWSN is inspired by three important ideas in SDN, namely, the separation of controller and data planes, standardized application interfaces and centralized management. As presented in Fig. 2, the differences between our SDWSN and SDN are:

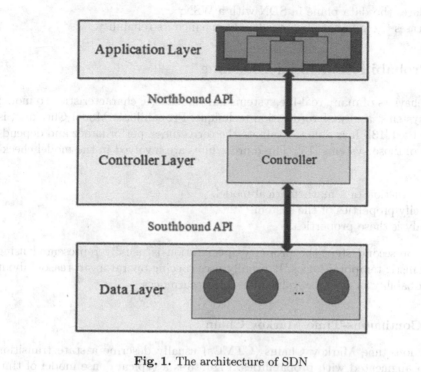

Fig. 1. The architecture of SDN

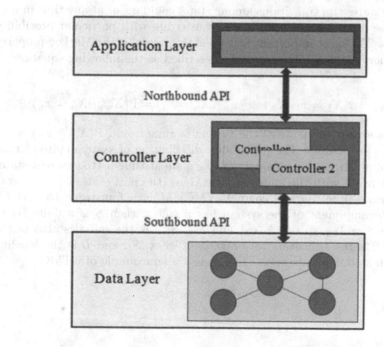

Fig. 2. The architecture of SDWSN

- Replaces the data plane in SDN with a WSN;
- Increases the number of controllers to enhance its reliability.

3   Probabilistic Model Checking

The behaviors of many real-life systems have stochastic characteristic. To model these systems, a formal verification technique, Probabilistic Model Checking, is widely used [13]. It is able to evaluate the correctness, performance and dependability of those systems [13]. Three procedures are involved in the model checking:

- Construction of a mathematical model.
- Specify properties of the system.
- Analysis these properties.

In the second step, the property specification is usually represented using probabilistic temporal logics. It usually uses specific operators to reason about the probability or Boolean value of events occurrences.

3.1   Continuous-Time Markov Chain

Continuous-time Markov Chains (CTMCs) usually describe a state transition system augmented with probabilities. Figure 3 is a typical dense model of time with discrete states. Its transitions among states could occur at any time unit in CTMC. These arrows present the state transferring with particular possibility in the system. The most important feature of this model is the Markov property that is "memorylessness". This could be described as the following equation:

$$P(X_{n+1} = x | X_1 = x_1, X_2 = x_2, \ldots, X_n = x_n) = P(X_{n+1} | X_n = x_n)$$

where X_n stands for the state of the system at time n and $P(X_n = x_n)$ is the corresponding probability. The probability distribution of system state at time $n + 1$ (the future state) depends only on the state at time n (the current state) and does not relate with the states at time $(1, n)$ (the past states).

A continuous-time Markov chain (CTMC) M is a four-tuple (S, s, R, L). Assume each component of the system has n states, then S is a finite set of states where $S = \{s_0, s_1, s_2, \ldots, s_n\}$ and $s \in S$; s is the initial status of the component; R is the transition rate matrix of $|S| \times |S|$, and L is the labeling function with atomic propositions [2]. Figure 3 is an example of CTMC:

$$S = \{s_0, s_1, s_2, s_3, s_4\};$$

$$s = s_0;$$

$$R = \begin{bmatrix} 0 & 0.5 & 0.2 & 0.3 & 0 & 0 \\ 0 & 0 & 0 & 0 & 0.7 & 0 \\ 0 & 0 & 0 & 0 & 0 & 0 \\ 0 & 0 & 0 & 0 & 0 & 0.2 \\ 0 & 0 & 0 & 0 & 0 & 0.2 \end{bmatrix};$$

$L(s_0) = \{OK\}, L(s_1) = L(s_3) = L(s_4) = \{\}, L(s_2) = \{Failed\}, L(s_5) = \{Waiting\}.$

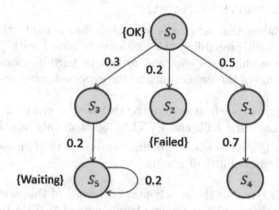

Fig. 3. CTMC example.

Time in CTMC is generally modeled using exponential distribution [16]. For a continuous random variable X, if its density function is given by:

$$f(t) = \begin{cases} \lambda e^{-\lambda t}, & \text{if } t > 0 \\ 0, & \text{otherwise} \end{cases}$$

where λ is the rate, X is exponential with the parameter $\lambda > 0$. In this system, all delays and repair time are assumed as X here, therefore, the property $P(X \leq t)$ can be computed by

$$F(t) = P(X \leq t) = \int_0^t \lambda e^{-\lambda t} dx$$

and the property $P(X > t)$ is $e^{-\lambda t}$.

3.2 Continuous Stochastic Logic

Continuous stochastic logic (CSL) is used for specifying properties of CTMCs. It is defined by two types of syntaxes: state formula (Φ) and path formula (Ψ). CSL allows many reliability related properties to be specified such as the reachability property [13]. This usually outputs a failure rate where the transaction of the system modeled by CTMC finally shut down, or a success rate. In this project, we mainly focus on the failure rate of the system.

3.3 PRISM

PRISM is a popular probabilistic model checking tool. It now mainly supports four types of probabilistic model: discrete-time Markov chains (DTMCs), continuous-time Markov chains (CTMCs), Markov decision processes (MDPs) and Probabilistic timed automata (PTAs) [8].

The model in PRISM is specified using the PRISM modeling language. Each System is described as modules. Each module contains its local variables which is used to indicate the possible states. The status transition behavior in modules is represented in command. For example:

```
[] x=0 -> 0.8:(x'=0) + 0.2:(x'=1);
```

This command claims that when the variable x has value 0, the update state will remain at 0 with probability 0.8 and change into 1 with probability 0.2. Additionally, the module uses cost and reward to permit reasoning about the quantitative measure of the system such as the expected power consumption and expected time.

PRISM's property specification language is based on temporal logics, namely, PCTL, CSL, probabilistic LTL and PCTL* [13]. It mainly has three operators:

- P operator: this is used to reason about the probability of an event occurrence, e.g. what is the probability of $y = 6$:
  `P=? [y = 6]`.
- S operator: this refers to the steady-state behavior of the model, e.g. what is the long-run probability of the queue being more than 75 % full:
  `S=? [q > 0.75]`.
- R operator: this is used for reasoning about the expected cost of the model. For example: what is the expected number of lost requests within 15.5 time units
  `R=? [C <= 15.5]`.

4   Model Checking SDWSN

This section illustrates a SDWSN formal model and presents its implementation details.

4.1 Formal Model of a Typical SDWSN

PRISM is used to analyze SDWSN reliability. To achieve this, the general SDN architecture shown in the Fig. 1 is modified into the structure presented in the Fig. 2. The improved SDWSN has following features:

- Application Layer: Because this paper does not consider the effects of applications, all of them are regarded as a whole with very low failure rate.
- Controller Layer: If the controller fails, the system will shut down. Hence, the modified structure adds one controller to increase the reliability.
- WSN Layer: This consists of several sensors to collect and measure information and other physical quantities. Similar with control plane, if all sensors fail, then controller will shut down the system.

4.2 Implementation

In the formal model, two modules (controller layer and WSN layer) are realized. The detail structures of modules are as follows:

- Module 1 - Controller layer: Two controllers are c_1 and c_2. The states are: both two controllers are working ($m = 2$), only one is working ($m = 1$) and fail ($m = 0$). Each controller has the same failure rate λ_c. The transition rate of the controller layer whose initial status is $m = 2$ to the state $m - 1$ is $m \times \lambda_c$.
- Module 2 - WSN layer: At least two sensors are considered in this module. Additionally, the number of sensors can be manually changed. Each sensor has the same failure probability λ_a.

To verify the dependability of this model, the following failure types are considered:

- Controller layer fails if none of controllers is working.
- WSN fails if none of sensors is working.
- System fails if the controller layer fails.
- System fails if the number of working sensors is less than the lower bound.
- System fails if the skipping cycle is greater than the upper limit. (The controller can automatically skip the current operation cycle when the application or WSN layer is abnormal, then a Max_COUNT is set to limit the skipping.)

5 Results

The main results are presented in this section. The factors of influencing the reliability of SDWSN are also analyzed.

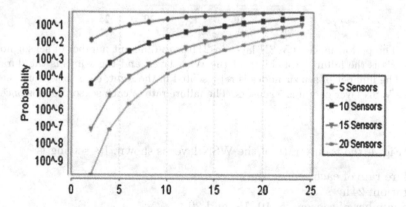

Fig. 4. The performance of WSN layer in 24 h with different number of sensor nodes. The y-axis is the failure probability of the WSN layer; and the x-axis is the duration (24 h). The number of sensor nodes is represented in the right; e.g., "5 sensors" means that the WSN layer contains 5 sensors. The failure rate of each sensor node is once a day.

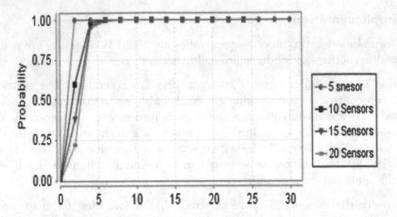

Fig. 5. The performance of WSN layer in 30 days with different number of sensor nodes. The y-axis is the failure probability of the WSN layer; and the x-axis is the duration (30 days). The number of sensor nodes is represented in the right; e.g., "5 sensors" means that the WSN layer contains 5 sensors. The failure rate of each sensor node is once a day.

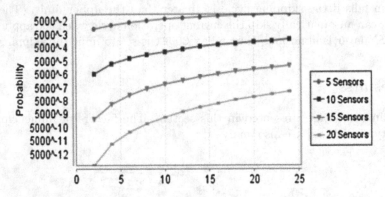

Fig. 6. The performance of WSN layer in 24 h with different number of sensor nodes. The y-axis is the failure probability of the WSN layer; and the x-axis is the duration (24 h). The number of sensor nodes is represented in the right; e.g., "5 sensors" means that the WSN layer contains 5 sensors. The failure rate of each sensor node is once per 30 days.

In Fig. 4, the failure rate of the WSN layer is shown. Its setting is:

- Failure rate of each sensor: once per day;
- Duration: 24 h;
- The number of sensors: 5, 10, 15, and 20.

As shown in Fig. 4, increasing the number of sensors can decrease the failure probability of the entire WSN layer. However, if we set the duration to 30 days, this strategy will be challenged. The result is shown in Fig. 5. From this figure, we can see that the failure probability of the WSN plane is 100 % after 4 days.

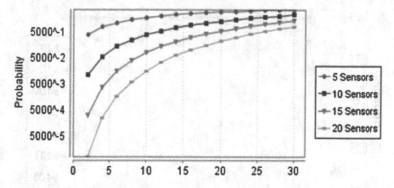

Fig. 7. The performance of WSN layer in 30 days with different number of sensor nodes. The y-axis is the failure probability of the WSN layer; and the x-axis is the duration (30 days). The number of sensor nodes is represented in the right; e.g., "5 sensors" means that the WSN layer contains 5 sensors. The failure rate of each sensor node is once per 30 days.

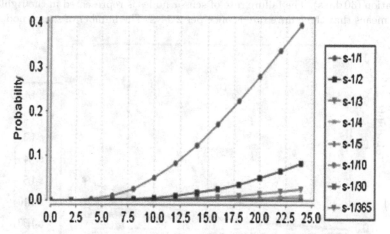

Fig. 8. The performance of WSN layer in 24 h with different failure rate of individual sensor nodes. The y-axis is the failure probability of the WSN layer; and the x-axis is the duration (24 h). The failure rate of sensor nodes is represented in the right; e.g., "s-1/2" means that the failure rate is once per 2 days. The number of sensor nodes is 5.

This suggests that when the failure probability of individual sensor node is high (once per day), increasing the number sensor nodes is not a good way of enhancing the reliability of the WSN layer.

In Figs. 6 and 7, the failure rate of each sensor in the WSN plane is changed to once per month. Compared with Figs. 4 and 5, the failure probability of the WSN plane is significantly lower. This depicts that decreasing the failure rate of each sensor node can decrease the failure probability of the sensor layer.

In order to study the relation between the sensor node failure rate and the WSN layer failure rate, we did several new experiments. In Figs. 8 and 9, we can

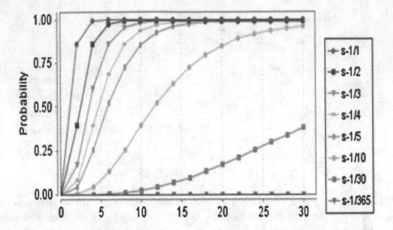

Fig. 9. The performance of WSN layer in 30 days with different failure rate of individual sensor nodes. The y-axis is the failure probability of the WSN layer; and the x-axis is the duration (30 days). The failure rate of sensor nodes is represented in the right; e.g., "s-1/2" means that the failure rate is once per 2 days. The number of sensor nodes is 5.

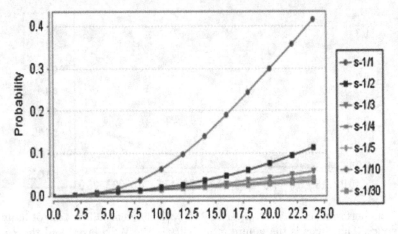

Fig. 10. The performance of the whole system (SDWSN) in 24 h with different failure rate of individual sensor nodes. The y-axis is the failure probability of the system; and the x-axis is the duration (24 h). The failure rate of sensor nodes is represented in the right; e.g., "s-1/2" means that the failure rate is once per 2 days. The number of sensor nodes is 5. The failure rate of the controller is?

see that lower sensor node failure rate is corresponding to lower WSN failure rate. In Figs. 10 and 11, we can see that lower sensor node failure rate can also enhance the system (SDWSN) reliability given a fixed controller failure rate.

Except the sensor nodes, the system reliability is also related to controllers' status. We have the following findings:

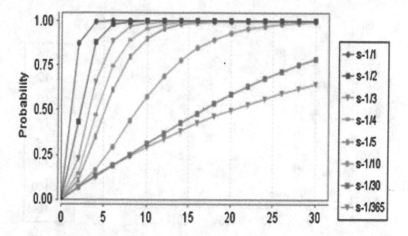

Fig. 11. The performance of the whole system (SDWSN) in 30 days with different failure rate of individual sensor nodes. The y-axis is the failure probability of the system; and the x-axis is the duration (30 days). The failure rate of sensor nodes is represented in the right; e.g., "s-1/2" means that the failure rate is once per 2 days. The number of sensor nodes is 5. The failure rate of the controller is once per 30 days

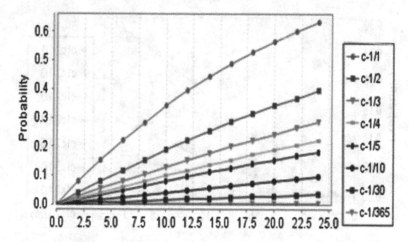

Fig. 12. The performance of the whole system (SDWSN) in 24 h with different failure rate of one controller. The y-axis is the failure probability of the system; and the x-axis is the duration (24 h). The failure rate of the single controller is represented in the right; e.g., "c-1/2" means that the failure rate is once per 2 days. The number of sensor nodes is 5. The failure rate of the sensor node is once per 30 days.

- From Figs. 12 and 13, we can see that reducing the failure rate of each controller can enhance the system reliability.
- From Figs. 12 and 14, we can see that increasing the number of controllers can also enhance the system reliability.

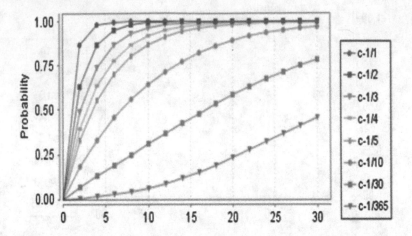

Fig. 13. The performance of the whole system (SDWSN) in 30 days with different failure rate of one controller. The y-axis is the failure probability of the system; and the x-axis is the duration (30 days). The failure rate of the single controller is represented in the right; e.g., "c-1/2" means that the failure rate is once per 2 days. The number of sensor nodes is 5. The failure rate of the sensor node is once per 30 days.

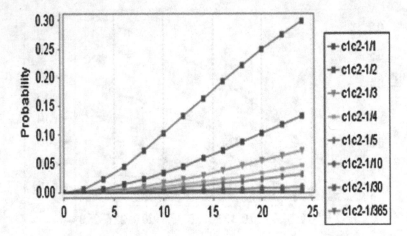

Fig. 14. The performance of the whole system (SDWSN) in 24 h with different failure rate of the two controllers. The y-axis is the failure probability of the system; and the x-axis is the duration (24 h). The failure rate of the two controllers is represented in the right; e.g., "c1c2-1/2" means that the failure rate is once per 2 days. The number of sensor nodes is 5. The failure rate of the sensor node is once per 30 days.

6   Conclusion

In this paper, to verify the reliability of SDWSN, a probabilistic model has been realized based on CTMC using PRISM. After experiments, the results illustrate the impacts of controllers and sensors on the network performance.

Firstly, increasing the number of controllers and sensors can enhance the system reliability. Secondly, the lower failure rate of controllers and sensors makes the network more dependable. Moreover, these two factors are more cognizable in long term.

Nevertheless, the limitation of the assumptions inevitably results in obstacles to find more potential correlation among components in SDWSN. For example, the relationship between the controller and sensor is omitted and this will be considered in the further research.

Acknowledgment. This work has been supported in part by the XJTLU RDF140243, by the Natural Science Foundation of Jiangsu Province under Grant BK20150376, and by the Suzhou Science and Technology Development Plan under Grant SYG201516.

References

1. Ali, S.T., Sivaraman, V., Radford, A., Jha, S.: A survey of securing networks using software defined networking. IEEE Trans. Reliab. **64**(3), 1086–1097 (2015)
2. Aziz, A., Sanwal, K., Singhal, V., Brayton, R.: Model-checking continuous-time Markov chains. ACM Trans. Comput. Logic (TOCL) **1**(1), 162–170 (2000)
3. De Mil, P., Jooris, B., Tytgat, L., Hoebeke, J., Moerman, I., Demeester, P.: snap-Mac: a generic MAC/PHY architecture enabling flexible MAC design. Ad Hoc Netw. **17**, 37–59 (2014)
4. Ding, A.Y., Crowcroft, J., Tarkoma, S., Flinck, H.: Software defined networking for security enhancement in wireless mobile networks. Comput. Netw. **66**, 94–101 (2014)
5. Granelli, F., Gebremariam, A.A., Usman, M., Cugini, F., Stamati, V., Alitska, M., Chatzimisios, P.: Software defined, virtualized wireless access in future wireless networks: scenarios and standards. IEEE Commun. Mag. **53**(6), 26–34 (2015)
6. Han, Z.J., Ren, W.: A novel wireless sensor networks structure based on the SDN. Int. J. Distrib. Sens. Netw. **2014** (2014)
7. Harrop, P., Das, R.: Wireless sensor networks 2010-2020. Networks **2010**, 2020 (2010)
8. Hinton, A., Kwiatkowska, M., Norman, G., Parker, D.: PRISM: a tool for automatic verification of probabilistic systems. In: Hermanns, H., Palsberg, J. (eds.) TACAS 2006. LNCS, vol. 3920, pp. 441–444. Springer, Heidelberg (2006)
9. Hunkeler, U., Lombriser, C., Truong, H.L., Weiss, B.: A case for centrally controlled wireless sensor networks. Comput. Netw. **57**(6), 1425–1442 (2013)
10. Jacobsson, M., Orfanidis, C.: Using software-defined networking principles for wireless sensor networks. In: 11th Swedish National Computer Networking Workshop (SNCNW), 28–29 May 2015, Karlstad, Sweden (2015)
11. Jagadeesan, N.A., Krishnamachari, B.: Software-defined networking paradigms in wireless networks: a survey. ACM Comput. Surv. (CSUR) **47**(2), 27 (2014)
12. Kirkpatrick, K.: Software-defined networking. Commun. ACM **56**(9), 16–19 (2013)
13. Kwiatkowska, M., Norman, G., Parker, D.: Probabilistic model checking in practice: case studies with PRISM. ACM SIGMETRICS Perform. Eval. Rev. **32**(4), 16–21 (2005)
14. Lin, Y.-D., Lin, P.-C., Yeh, C.-H., Wang, Y.-C., Lai, Y.-C.: An extended SDN architecture for network function virtualization with a case study on intrusion prevention. IEEE Netw. **29**(3), 48–53 (2015)

15. Luo, T., Tan, H.-P., Quek, T.Q.S.: Sensor openflow: enabling software-defined wireless sensor networks. IEEE Commun. Lett. **16**(11), 1896–1899 (2012)
16. Menshikov, M., Petritis, D.: Explosion, implosion, and moments of passage times for continuous-time Markov chains: a semimartingale approach. Stoch. Process. Appl. **124**(7), 2388–2414 (2014)
17. Oualhaj, O.A., Kobbane, A., Sabir, E., Ben-othman, J., Erradi, M.: A ferry-assisted solution for forwarding function in wireless sensor networks. Pervasive Mob. Comput. (2015)
18. Xu, R., Huang, X., Zhang, J., Lu, Y., Wu, G., Yan, Z.: Software defined intelligent building. Int. J. Inf. Secur. Priv. (2016)
19. Steiner, R.V., Mück, T.R., Fröhlich, A.A.: C-MAC: a configurable medium access control protocol for sensor networks. In: 2010 IEEE Sensors, pp. 845–848. IEEE (2010)
20. Van Dam, T., Langendoen, K.: An adaptive energy-efficient MAC protocol for wireless sensor networks. In: Proceedings of the 1st International Conference on Embedded Networked Sensor Systems, pp. 171–180. ACM (2003)
21. Wood, T., Ramakrishnan, K.K., Hwang, J., Liu, G., Zhang, W.: Toward a software-based network: integrating software defined networking and network function virtualization. IEEE Netw. **29**(3), 36–41 (2015)
22. Xie, J., Guo, D., Zhiyao, H., Ting, Q., Lv, P.: Control plane of software defined networks: a survey. Comput. Commun. **67**, 1–10 (2015)
23. Lu, Y., Huang, X., Huang, B., Xu, W., Zhang, Q., Xu, R., Liu, D.: A study on the reliability of software defined wireless sensor network. In: 2015 IEEE International Conference on Smart City (IEEE Smart City 2015), Chengdu, China, 19–21 December 2015

A Path-Line-Based Modeling for Fishing Prediction

Koji Koyamada[1]([⊠]), Katsumi Konishi[2], Naohisa Sakamoto[1],
and Marohito Takami[1]

[1] Academic Center for Computing and Media Studies,
Kyoto University, Kyoto, Japan
koyamada.koji.3w@kyoto-u.ac.jp
[2] Department of Computer Science,
Kogakuin University, Tokyo, Japan

Abstract. The development of a predictive model for fishing grounds is required due to climate change and global warming to maintain steady fish catches. In general, the fish catches are modeled using environment variables in the time-varying ocean simulation dataset. The current modeling techniques consider ocean simulation dataset only on the fishing day, which is too simple to explore a causal relationship between the catches and environment variables. To solve this problem, we use environmental variables, such as water temperature, velocity and salinity, sampled on flow path-lines, which terminate at fishing locations, and we employ a standard least squares method with the L_0 norm regularization technique to construct an accurate fishing ground model. We apply the approach to construct a predictive model of neon flying squid to confirm the effectiveness by demonstrating the model accuracy distribution along the path-lines.

Keywords: Fishing ground modeling · L_0 norm regularization · Visualization

1 Introductions

Global warming is causing climate change, which increases the variance in environmental variables, such as temperature, salinity and velocity. Climate change results in a change of locations of the habitats of fish because the habitats of fish are always looking for a favorable environment. For fishery grounds modeling, such an environment has been represented as a function of environmental variables. If the number of the habitats and the environmental variables can be measured, we can construct a predictive model for the population.

There is a strong need for the development of a predictive model for fishing to maintain steady fish catches. Fishing is under greater risk due to global warming because the fisherman's tacit knowledge may be useless. To avoid wasting fuel oil in fishing, the fishermen understand that they should have a fishery prediction technique. The fishermen asked the academic community to develop a highly accurate predictive model for their fishery. They recorded fish catches and made the results open to the community.

© Springer Science+Business Media Singapore 2016
S.Y. Ohn and S.D. Chi (Eds.): AsiaSim 2015, CCIS 603, pp. 79–88, 2016.
DOI: 10.1007/978-981-10-2158-9_7

Recently, fishermen have recorded their fish catches when they are on their fish boats for future reference. The catches are represented using a unit named the catch per unit effort (CPUE). In fisheries, the CPUE is an indirect measure of the abundance of a target species. Changes in the catch per unit effort are inferred to signify changes to the true abundance of the target species. The CPUE is recorded with the location at which the fishing is conducted. The location includes the latitude and longitude of the fishing boat but not the depth at which the fish are caught.

In fishery ground prediction, the CPUE is generally represented as a function of environment variables. It is difficult to measure sufficient variable data. High performance computing facilities enable us to calculate a highly accurate environmental model that includes the temperature, salinity and velocity. With down-scaling and data assimilation techniques, we can obtain highly accurate environmental variables with sufficient resolution around possible fishing grounds.

Current modeling techniques consider an ocean simulation dataset only on the fishing day. With such a technique, which does not use time-varying datasets, it is difficult to explore the causal relationship between the CPUE and environment variables, even if their fitness has been maximized. The food chain is important when we estimate fishery prediction. It originates from nutrient salts that precipitate at the bottom of the sea. When the salts move upwards where the sunlight arrives along an upwelling current, phytoplankton can reproduce. Then, zooplankton and fish eat the phytoplankton and the zooplankton, respectively. It is natural that we assume that these food chain components follow the sea water current. The phytoplankton and the zooplankton follow the ocean current passively in their moving.

In our research, we employ a standard least squares method with the L_0 norm regularization technique [3] to sparsely model the causal relationship because we have a large number of possible explanatory variables considering a time-varying ocean simulation dataset. The technique is a type of signal processing technique for efficiently acquiring and reconstructing a signal by finding solutions to underdetermined linear systems based on the principle that, through optimization, the sparsity of a signal can be exploited to recover it from far fewer samples than required by the Shannon-Nyquist sampling theorem.

The remainder of the paper is organized as the follows. In Sect. 2, we describe the related work on fishery grounds predictive modeling. In Sect. 3, we explain our visual analytics system, which supports the discovery of the food chains along a path-line in the ocean flow. In Sect. 4, we describe the L_0 norm regularization technique for sparse modeling. In Sect. 5, we apply the technique to construct a predictive fishery model for neon-flying fishes to confirm the effectiveness by comparing the proposed technique with the previous one.

2   Related Work

Tian et al. developed two types of HSI models for predicting neon squid fish catches and compared the models [1]. The HSI model represents an index that indicates whether an animal feels comfortable by using a value from 0.0 to 1.0. The index is defined by the United States Environmental Preservation Agency (EPA). In this HSI model, a Suitability Index (SI) is first constructed so that the CPUE can be described using a single environmental variable, such as temperature, salinity or velocity. We call this method Tian's method. For example, when we consider a sea animal, we select one of the temperature, salinity and velocity variables and construct a function of the CPUE with the selected variable. To develop an HSI model, we integrate several SI indexes into a single HSI model using a mathematical method, such as the geometric mean.

Xinjun et al. proposed an HSI model in which the CPUE is explained using a set of weighted summations of the environmental variables and evaluated the effect caused by the weighting methods [2]. We call this method Xinjun's method. These previous models considered only the correlation between the CPUE and the environmental variables but did not consider any causal relationship in the food chain.

Koyamada et al. proposed a modeling technique that employs environmental variables, such as water temperature, velocity and salinity, sampled on flow path-lines that terminate at fishing locations [6]. Although the results indicated that the accuracy of the model was improved when it considered the variables sampled along the path-lines, they did not determine how many sampling points along the path-line are required for effective modeling.

3 Visual Analytics System for a Path-Line-Based Sampling

For the exploration of the causal relationship in the food chain, we focus on a path-line that represents a curved line whose tangent is parallel to a velocity vector at a spatio-temporal point. In this research, we assume that the food chain components, such as plankton and fish, move along the ocean current. Although the number of the target fish that the fishermen try to catch may eat small fishes can be measured by the CPUE, it is difficult to measure those of the other components. It is naturally inferred that each can have a set of environmental variables and specific value ranges.

Along a path-line that passes through a fishing location, we sample environmental variables, such as the water temperature, velocity and salinity. We regard these variables as the factors that can be regarded as the occurrences of the food chain components. In other words, these variables can be explained by the latent variables of the occurrences.

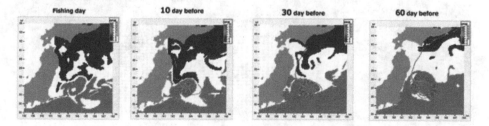

Fig. 1. Fishing ground points and the corresponding path-lines. (Color figure online)

We developed a visual analytic system to confirm that a path-line starts at Oyashio region, which can provide nutrient salts, and terminates at a CPUE point. Figure 1 shows the CPUE points and the corresponding path-lines that pass through these points. The left image shows a visualization result of the fishing day. The second, third and fourth images show the results of the previous ten, thirty, and sixty days, respectively. The blue and orange regions represent Oyashio and Kuroshio regions, respectively. If the food chain is conducted along the path-line, the expert can divide it into multiple segments that are related to food chain components. Because the first component is phytoplankton, which often reproduces by the upwelling current, it is preferable to identify a location where the reproduction starts to occur. We asked domain experts to use and to evaluate the visual analytic system of whether the food chain can be identified along the path-line. The domain experts agreed to its effectiveness and suggested that the existence of vortex flow may be related to the formation of the food chain.

4   Method

We develop a fishing ground model using an ocean simulation dataset and a fish catch dataset called CPUE. The simulation dataset is generated from a data assimilation system for the Northwest Pacific Ocean. The values of the environmental variables were obtained from an ocean simulation based on the 3D-VAR data assimilation product MOVE (MRI Multivariate Ocean Variational Estimation). The simulation generated values of temperature (T), salinity (S), and current velocity (U, V, W). Thus, the number of variable types is five. Because the grid size is 0.1° between 43° N and

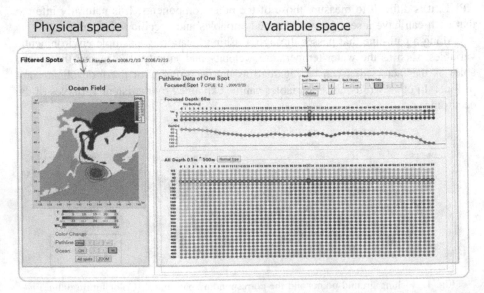

Fig. 2. Path-lines that terminate at a single CPUE point with varying depth values. (Color figure online)

35° S and 141° W and 147° E in the North-West Pacific Ocean, the grid resolution is 90 × 70 in the horizontal plane. We calculate the path-lines whose terminating points are located at the CPUE points. In the path-line, environmental variables, such as water temperature, velocity and salinity, are sampled [5]. Figure 2 shows the path-lines (left) and the sampled variables on the lines (right). In the lower right of the figure, several path-lines for a single CPUE point are displayed with the depth values changed.

In the fishing ground model, we assume that the fish catch is explained using environmental variables sampled along path-lines that are terminated at the CPUE points. We regard the depths of the CPUE points as unknown variables because there is no information on the depths where the fish were caught. Thus, we have 16,200 ($= 5 \times 54 \times 60$) variables at each CPUE point because the depth and time resolutions are 54 and 60, respectively. In our research, we consider two methods, Tian's and Xinjun's methods, to confirm the effectiveness in which time series data sets are employed for constructing the HSI model.

4.1 Xinjun's method

In Xinjun's method, we assume that the log CPUE can be described as a type of generalized linear model as follows:

$$\log CPUE = \sum_{k=1}^{60} \sum_{j=1}^{54} \left(\theta_{T_{k,j}} \log T_{k,j} + \theta_{S_{k,j}} \log S_{k,j} + \theta_{U_{k,j}} \log U_{k,j} + \theta_{V_{k,j}} \log V_{k,j} + \theta_{W_{k,j}} \log W_{k,J} \right),$$

$$(1)$$

where $T_{k,j}$, $S_{k,j}$, $U_{k,j}$, V and $W_{k,j}$ denote the temperature, salinity and velocity components of sea water at depth j and time step k, respectively, and $\theta_{k,j} = [\theta_{T_{k,j}} \theta_{S_{k,j}} \theta_{U_{k,j}} \theta_{V_{k,j}} \theta_{W_{k,j}}]^T$ are design parameters. To simplify the problem, the value of log $CPUE$ is roughly quantized to be in $\{1,2\}$, that is, in (1), we replace the value with $q_a(\log CPUE)$, where q_a is a quantization function defined by

$$q_a(x) = \begin{cases} 1 & \text{if } x < a \\ 2 & \text{if } x > a \end{cases}.$$

This paper proposes the following HSI model,

$$q_a(\log CPUE) = \sum_{k=1}^{60} \sum_{j=1}^{54} \left(\theta_{T_{k,j}} \log T_{k,j} + \theta_{S_{k,j}} \log S_{k,j} + \theta_{U_{k,j}} \log U_{k,j} + \theta_{V_{k,j}} \log V_{k,j} + \theta_{W_{k,j}} \log W_{k,j} \right)$$

$$(2)$$

In the numerical experiments of the next section, we use the average of log $CPUE$ as the threshold a.

Next, we focus on a scheme to estimate the parameters $\theta_{k,j}$. Let us consider N CPUE points $CPUE^{(i)}$ and their environmental variables $T_{k,j}^{(i)}$, $S_{k,j}^{(i)}$, $V_{k,j}^{(i)}$, $U_{k,j}^{(i)}$ and $W_{k,j}^{(i)}$ for

$i = 1, 2, \ldots, N$. The parameters can be obtained using a standard least squares method, and we use L_0 norm regularization to avoid overfitting and propose the following problem,

$$\text{minimize } \|y - A\theta\|^2 + \lambda\|\theta\|_0, \tag{3}$$

where $\|\cdot\|_0$ denotes the L_0 norm of a vector, that is, the number of non-zero entries of a vector, $y \in \mathcal{R}^N$ is a constant vector whose i th entries are $CPUE^{(i)}$, $A \in \mathcal{R}^{N \times 16200}$ is a constant matrix consisting of $\log T_{k,j}^{(i)}$, $\log S_{k,j}^{(i)}$, $\log V_{k,j}^{(i)}$, $\log U_{k,j}^{(i)}$ and $\log W_{k,j}^{(i)}$ as follows,

$$a_{kj}^i = \left[\log T_{kj}^{(i)} \log S_{kj}^{(i)} \log V_{kj}^{(i)} \log U_{kj}^{(i)} \log W_{kj}^{(i)}\right],$$

$$A = \begin{bmatrix} a_{1,1}^1 & a_{1,2}^1 & \cdots & a_{1,54}^1 & a_{2,1}^1 & \cdots & a_{60,54}^1 \\ a_{1,2}^2 & a_{1,2}^{12} & \cdots & a_{1,54}^2 & a_{2,1}^2 & \cdots & a_{60,54}^2 \\ \vdots & \vdots & & \vdots & \vdots & & \vdots \\ a_{1,2}^N & a_{1,2}^N & \cdots & a_{1,54}^N & a_{2,1}^N & \cdots & a_{60,54}^N \end{bmatrix},$$

and $\theta \in \mathcal{R}^{16200}$ is a design variable vector. To solve this problem, we apply the iterative reweighted least squares (IRLS) algorithm [4]. In IRLS, the solution at the $(t+1)$ th iteration is obtained as

$$\theta^{(t+1)} = \arg\min_{\theta}\|y - A\theta\|^2 + \lambda\|W^{(t)}\theta\|^2, \tag{4}$$

where $W^{(t)}$ is a diagonal matrix whose diagonal elements $W_{ii}^{(t)}$ are calculated using the optimal solution of the previous iteration as follows,

$$W_{ii}^{(t)} = \frac{1}{\left|\theta_i^{(k)}\right| + \varepsilon}, \tag{5}$$

and ε is a small constant. The empirical results show that we can obtain the optimal solution of (3) by iterating (4) and (5) until convergence.

IRLS algorithm for (3)

Step 1.   Set $W_{ii}^{(0)}$ to be the identity matrix for $i = 1, 2, \cdots, 16200$.

Step 2.   Set $t \leftarrow 1$.

Step 3.   Repeat Step 3-1 to Step 3-3 until $\left\|\theta^{(t+1)} - \theta^{(t)}\right\| < tol$

Step 3-1.   Calculate $\theta^{(t+1)}$ using (4).

Step 3-2.   Calculate $W_{ii}^{(t+1)}$ for $i = 1, 2, \cdots, 16200$ using (5).

Step 3-3. $t \leftarrow t + 1$

4.2   Tian's Method

In Tian's method, we assume that the log CPUE, that is the SI model, can be described as a type of generalized linear model for each variable as follows

$$SI_X = \log CPUE = \sum_{k=1}^{60}\sum_{j=1}^{54}\theta_{X_{k,j}}\log X_{k,j}, \tag{6}$$

where X denotes S, T, V, U or W. Then, using the quantization function defined in the previous section, this method defines the following HSI model:

$$HSI = q_a\left(\frac{SI_S + SI_T + SI_V + SI_U + SI_W}{5}\right). \tag{7}$$

Next, we focus on a scheme to estimate the parameters $\theta_{X_{k,j}}$. Let us consider N CPUE points $CPUE^{(i)}$ and their environmental variables $X_{k,j}^{(i)}$ for $i = 1, 2, \ldots, N$, where X denotes S, T, V, U or W. The parameters can be obtained using Eq. 3, in which $A \in \mathcal{R}^{N \times 3240}$ is a constant matrix consisting of $z_{k,j}^i = \log X_{k,j}^{(i)}$ as follows:

$$A = \begin{bmatrix} z_{1,1}^{(1)} & z_{1,2}^{(1)} & \cdots & z_{1,54}^{(1)} & z_{2,1}^{(1)} & \cdots & z_{60,54}^{(1)} \\ z_{1,1}^{(2)} & z_{1,2}^{(2)} & \cdots & z_{1,54}^{(2)} & z_{2,1}^{(2)} & \cdots & z_{60,54}^{(2)} \\ \vdots & \vdots & & \vdots & \vdots & & \vdots \\ z_{1,1}^{(N)} & z_{1,2}^{(N)} & \cdots & z_{1,54}^{(N)} & z_{2,1}^{(N)} & \cdots & z_{60,54}^{(N)} \end{bmatrix}$$

and $\theta \in \mathcal{R}^{3240}$ is a design variable vector. To solve this problem, we apply the IRLS algorithm described in the previous section for each variable S, T, V, U or W.

5   Experimental Results and Discussion

In this research, an important point is how sparsely the CPUE can be modeled, that is, how much the number of variables is minimized. To examine the method proposed in the previous section, we perform some numerical experiments. In all experiments, we use $\varepsilon = 10^{-8}$ and $tol = 10^{-5}$. Because the optimal solution of (3) depends on the value of λ, we increase it from 0.01 to 1 by 0.1 and select the best λ.

To show the efficiency of the proposed method, we compare the proposed method with the previous method, in which a time-varying dataset is not employed. In Xinjun's method, the CPUE can be described by leaving terms in which k equals 1 in Eq. 6.

$$\log CPUE = \sum_{j=1}^{54}\left(\theta_{T_i}\log T_{1,i} + \theta_{S_i}\log S_{1,i} + \theta_{T_i}\log U_{1,i} + \theta_{T_i}\log V_{1,i} + \theta_{T_i}\log W_{1,i}\right),$$

In Tian's method, we assume that the log CPUE, that is, the SI model, can be described in the same way.

$$SI_X = \log CPUE = \sum_{j=1}^{54} \theta_{X_{k,j}} \log X_{k,j}$$

where X denotes S, T, V, U or W.

Here, the parameters are estimated by a least squares method. We examine these methods using fish cash data of January 2006. The parameters are estimated using 45 CPUE points of January 1 to 10, and we examine the accuracy rate of the estimated 137 CPUE points from January 11 to 31. Because the values of $\log CPUE$ should be 1 or 2, we round them into 1 or 2 and calculate the percentage of correct answers.

Table 1 shows the results that indicate that the accuracy of the model is improved when it considers the variables that are sampled along the path-lines.

The purpose of this research is to demonstrate the advantage that the environmental variables sampled along path-lines can explain the CPUE points that are measured at the terminating points. Numerical experiments indicate that the value of CPUE depends highly on the velocity of water. We showed the effectiveness by comparing the proposed model with the previous one, which does not consider the path-line sampling and the efficiency of the L_0 norm regularization to select important environmental variables.

Table 1. Percentage of correct answers.

Proposed method with best λ	75.91 %
Proposed method with $\lambda = 0$	49.64 %
Previous method using Eq. 6	54.95 %

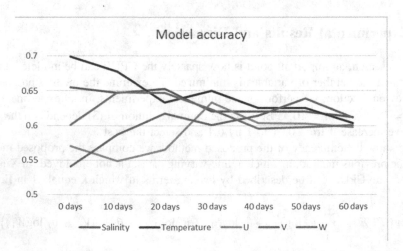

Fig. 3. Model accuracy distribution along the number of days. (Color figure online)

By considering a food-chain in the fishery, the model accuracy can be affected by the 60-day data fields at the maximum. To estimate the change of the accuracy of the models that are constructed using the 0-, 10-, 20-, 30-, 40-, 50- and 60-day data fields, we constructed the related prediction models. Because we employ five variables, which are the temperature, salinity, velocity vectors (U, V, W), we had five SI models, which were evaluated with respect to the accuracy. The HSI model is constructed by taking the geometrical average of the SI models. Figure 3 shows that the model accuracy cannot be improved by extending the number of days for the model construction. In many cases, the accuracy is maximized when 30-day data fields are used, which suggests that the data fields are not effective for more than 30 days.

6   Conclusion

To construct an accurate fishing ground model, we employed environmental variables, such as water temperature, velocity and salinity, sampled on flow path-lines that terminate at fishing locations and employ a standard least squares method with the L0 norm regularization technique. By comparing the developed model with a previous method that employs environmental variables only on the fishing day, we showed that the accuracy of the model is improved when it considers variables sampled along the path-lines. In addition, we calculate the model accuracy by changing the number of days in which the path-lines are traced back and understand that more than 30 days may not improve the model accuracy.

The food chain is governed by a physical law in which the sea current transports the environmental resources from the upstream regions to the downstream regions. In our model, we consider the treatment that reflects the physical law. In this research, we assume that the food chain is governed only by the sea current. In the future, we will consider physical transportation caused by the gradient vector flow of a scalar variable, such as salinity. We will also consider a vortex flow to explain a fish catch as the domain experts suggested after they experienced our visual analytics system. One piece of evidence shows the formation of twin vortices located near the fishing grounds [5]. The flow direction indicates that one of the two vortex centers attracts flow from the bottom of the ocean, which suggests that upwelling flow occurs near the vortex center and the upwelling may facilitate the food chain.

References

1. Tian, S., Chen, X., Chen, Y., Xu, L., Dai, X.: Evaluating habitat suitability indices derived from CPUE and fishing effort data for Ommatrephes bratramii in the northwestern Pacific Ocean. Fish. Res. **95**(2–3), 181–188 (2009)
2. Xinjun, C., Siquan, T., Bilin, L., Yong, C.: Modeling a habitat suitability index for the eastern fall cohort of Ommastrephes bartramii in the central North Pacific Ocean. Fish. Res. **29**(3), 493–504 (2011)
3. Daubechies, I., Devore, R., Fornasier, M., Gntrk, C.S.: Iteratively reweighted least squares minimization for sparse recovery. Comm. Pure Appl. Math **63**(1), 1–38 (2010)

4. Futami, K., Onoue, Y., Sakamoto, N., Koyamada, K.: Visualization of ocean current effects for fishing ground formation. J. Jpn Soc. Fluid Mech. **34**(2), 83–86 (2015)
5. Uenaka, T., Sakamoto, N., Koyamada, K.: Visual analysis of habitat suitability index model for predicting the locations of fishing grounds. In: Proceedings of IEEE Pacific Visualization 2014 (Visualization Notes), pp. 306–310 (2014)
6. Koyamada, K., Konishi, K., Sakamoto, N., Takami, M.: A path-line-based approach for developing a fishing ground model. In: Proceedings of AsiaSim 2015 (2015)

Simulation and Analysis

Numerical Simulation
of a Hamilton-Jacobi-Bellman Equation
for Optimal Management Strategy of Released
Plecoglossus Altivelis in River Systems

Yuta Yaegashi[1(✉)], Hidekazu Yoshioka[2], Koichi Unami[1],
and Masayuki Fujihara[1]

[1] Graduate School of Agriculture, Kyoto University, Kyoto, Japan
Yaegashi.yuta.54s@st.kyoto-u.ac.jp,
unami@adm.kais.kyoto-u.ac.jp,
fujihara@kais.kyoto-u.ac.jp
[2] Faculty of Life and Environmental Science,
Shimane University, Matsue, Japan
yoshih@life.shimane-u.ac.jp

Abstract. A stochastic differential equation model for population dynamics of released *Plecoglossus altivelis* (Ayu) in a river system subject to feeding damage by *Phalacrocorax carbo* (Great Cormorant) and fishing activity by human is proposed. A stochastic optimal control problem to maximize the sum of the cost of countermeasure to prevent the feeding damage and the benefit of harvesting the fish is formulated, which ultimately reduces to solving a Hamilton-Jacobi-Bellman equation. Application of a Petrov-Galerkin finite element scheme to the equation successfully computes the optimal management strategies for the population dynamics of *P. altivelis* in a real river system and ecological and economical indices to verify them.

Keywords: Inland fishery · *Plecoglossus altivelis* (Ayu) · *Phalacrocorax carbo* (Great Cormorant) · Population dynamics · Feeding damage · Stochastic optimal control · Hamilton-Jacobi-Bellman equation

1 Introduction

Plecoglossus altivelis (Ayu) is annual and diadromous fish species in Japan (Photo 1: left panel) [1]. *P. altivelis* accounts for 7.5 % (2,353 t) of total fish catch (31,264 t) of Japanese inland fisheries in 2013 [2]. Life history of *P. altivelis* is unique as reviewed in Tanaka et al. [3]. During autumn, adult fishes spawn eggs in the downstream reaches of their living river and die soon afterwards. Hatched larval fishes descend to coastal areas of the downstream water body of the river, which is a sea or a brackish lake in general, and grow up to juveniles with feeding on zooplanktons till the next spring. The grown fishes ascend the river toward its midstream and upstream reach where rock-attached algae, which are staple foods of *P. altivelis*, are available in riverbed. They feed on the algae to mature till the coming autumn when they descend the river.

© Springer Science+Business Media Singapore 2016
S.Y. Ohn and S.D. Chi (Eds.): AsiaSim 2015, CCIS 603, pp. 91–101, 2016.
DOI: 10.1007/978-981-10-2158-9_8

Photo. 1. Adult *P. altivelis* (left panel) and adult *P. carbo* (right panel). (The photo of *P. carbo* is taken from Photo by (c)Tomo.Yun http://www.yunphoto.net).

Literatures report that *P. altivelis* has significant influences on diversity of species in river ecosystems, such as aquatic insects, algae, and other fish species [4, 5]. Appropriately managing population of *P. altivelis* is therefore a key topic from both ecological and economical point of views.

Fish catch of *P. altivelis* in Japan has recently been rapidly decreasing, which is considered mainly due to decrease of its population [1]. A cause of the population decrease would include climate changes and manipulation of river environments such as installations of physical barriers like dams and weirs. Another not negligible cause is feeding damage by *Phalacrocorax carbo* (Great Cormorant), which is a piscivorous bird species that reigns as a top predator of river ecological systems in Japan (Photo. 2: right panel) [6]. Each individual of adult *P. carbo.* eats on average 0.5 (kg) of fish per day and has high ability to swim and to capture the preys in rivers. *P. carbo* lives in groups and creates colonies for nesting and spawning in riparian forest where there are less signs of human life. Total number of individuals of *P. carbo* in Japan was about three thousands during 1970 due to decline and decrease of habitats but has been recently increasing. Currently, total number of individuals of the bird in Japan is estimated to exceed more several tens of thousands [7]. Population increase of the bird has been causing significant feeding damage to inland fishery resources in Japan where *P. altivelis* is not an exception. Releasing juvenile fishes of *P. altivelis* in rivers under the initiative of local fishery cooperatives during every spring is currently a common way to artificially maintain population of *P. altivelis* subject to the predation from *P. carbo*. Fishery cooperatives and public bodies in Japan have been enthusiastically developing a number of direct and indirect countermeasures with different qualities to prevent feeding damage by *P. carbo* [6, 7], though decisive conclusion to determine the most optimal countermeasure has not been obtained yet. Theoretical investigation based on an appropriate mathematical model for population dynamics of *P. altivelis* considering predation from *P. carbo* would be useful for comparing performances of different countermeasures. However, only a few such researches have been conducted so far in Japan, which is the main motivation of this paper.

The purpose of this paper is to present a mathematical model for population dynamics of released *P. altivelis* in a river system subject to predation by *P. carbo* and fishing activity by human. An optimal control problem to maximize the sum of the profit by harvesting the fish and the cost of countermeasure to prevent the feeding damage is then presented. Finding the optimal management strategy of the population of *P. altivelis* reduces to solving a Hamilton-Jacobi-Bellman equation (HJBE), which is a time-backward degenerate parabolic nonlinear partial differential equation (PDE). The HJBE is numerically solved with a stable finite element scheme.

The rest of this paper is organized as follows. Section 2 presents the mathematical model used in this paper. Section 3 applies the mathematical model to optimal population management of released *P. altivelis* in Hii River, Shimane Prefecture, Japan. Section 4 concludes this paper and presents future perspectives of our research.

2 Mathematical Model

2.1 Stochastic Differential Equation Model

A stochastic process model for population dynamics of released *P. altivelis* in a habitat, which is a river system, is presented. Let $(0, T)$ with the terminal time $T(> 0)$ (day) be a period during which population dynamics of *P. altivelis* is considered. The time $t = 0$ (releasing time of juvenile *P. altivelis*) and $t = T$ (beginning of a closed season of fishing *P. altivelis*) are taken in summer and autumn, respectively within a year. The total biomass of *P. altivelis* in this river system at the time t (day) is denoted as X_t (kg). The initial condition X_0 (kg) at the time $t = 0$ (day) is assumed to be deterministic, which is considered to be valid for the case where most part of population of *P. altivelis* is introduced through intensive release events under the initiative of local fishery cooperatives [8]. Population growth of *P. altivelis* is assumed not to be prescribed by environmental capacity as in the usual ecological modelling [9] but rather by the growth curve of individuals, considering that the population of *P. altivelis* in such a case would not saturate in the river system. The governing Itô's stochastic differential equation (SDE) of the process X_t is proposed as [10]

$$dX_t = \left(a(t, X_t) - RX_t - k(u_t)X_t - \chi_{\{t \geq T_c\}}c_tX_t\right)dt + b(t, X_t)dB_t \qquad (1)$$

with the Verhulst-type coefficients arising from the growth curve of fishes as

$$a(t, x) = r(1 - K^{-1}x)x \qquad (2)$$

and

$$b(t, x) = \sigma x \qquad (3)$$

where B_t (day$^{1/2}$) is the 1-D standard Brownian motion, $K = mX_0$ (kg), m (−), r (1/day), and σ (1/day$^{1/2}$) are positive parameters, R (1/day) is the natural mortality rate (1/day), $k(u)(\geq 0)$ (1/day) is the predation pressure of *P. carbo* as a function of the control variable $(0 \leq)u(\leq 1)$ (−), which is effort to prevent the predation, $(0 \leq)c(\leq c_M)$ (1/day) is the fishing pressure, c_M is the maximum fishing pressure, $(0 <)T_c(< T)$ (day) is the opening time of catching *P. altivelis*, and χ_S is the characteristic function for generic set S. The term $b(t, X_t)dB_t$ describes temporal fluctuation of the population of $P.$ *altivelis* due to natural and artificial environmental changes in the habitat [11, 12]. A major difference with the present and conventional stochastic Verhulst models [9] is that the parameter K of the former depends on X_0 but that of the latter does not. Population dynamics of $P.$ *carbo* is not directly considered in the SDE (1), but its influences on the population dynamics of $P.$ *altivelis* are considered in the term $k(u_t)X_t$ to formulate a minimum mathematical model.

The generator $A^{u,c}$ for the coupled process $Y_t = (t, X_t)$ conditioned on $Y_s = (s, x)$ with $s < t$ for sufficiently regular $\varphi = \varphi(s, x)$ is expressed as [10]

$$A^{u,c}\varphi = \frac{\partial \varphi}{\partial s} + \left(a - Rx - k(u)x - \chi_{\{s \geq T_c\}}cx\right)\frac{\partial \varphi}{\partial x} + \frac{1}{2}b^2\frac{\partial^2 \varphi}{\partial x^2}. \qquad (4)$$

2.2 Optimal Control Problem

The variables u and c are taken as control variables in the model and are assumed to be Markov controls. The admissible ranges U and C of the control variables u and c are specified to be the compact sets as

$$U = \{u | 0 \leq u \leq 1\} \qquad (5)$$

and

$$C = \{c | 0 \leq c \leq c_M\}, \qquad (6)$$

respectively. The objective function to be maximized, which represents the total profit and is denoted as $v = v(s, T, X, u, c)$ (kg), is proposed as

$$v(s, T, X, u, c) = -\int_s^T f(u_t)dt + \int_s^T \chi_{\{t \geq T_c\}}c_t X_t dt \qquad (7)$$

where $f(\geq 0)$ (kg/day) with $f(0) = 0$ is an increasing function. The first and second terms in the right hand-side of (7) represent the minus of the total cost of operating a countermeasure to prevent the feeding damage by $P.$ *carbo* and the total amount of harvested $P.$ *altivelis* by human during the period (s, T), respectively. The coefficients k and f are set as

$$k(u) = k_0(1 - \alpha u) \qquad (8)$$

and

$$f(u) = \omega(\exp(\beta u) - 1), \qquad (9)$$

respectively where k_0 (1/day) is the predation pressure from the bird without any countermeasures, $(0 <)\alpha(\leq 1)$ modulates the effectiveness to decrease the predation pressure with the control u, $\beta(>0)$ ($-$) modulates the cost to prevent feeding damage, and ω (kg/day) is a parameter serving as a weight that determines the balances on the cost and the benefit in the objective function. The parameter α represents the effectiveness of the countermeasure, while β represents its inefficiency to prevent the feeding damage. Values of these parameters would depend on the countermeasures chosen [6, 7].

The goal of the optimal control problem is to find the optimal controls $u = u^*$ and $c = c^*$ that maximize the objective function v. The maximized objective function V (kg) is referred to as the value function, which is expressed as

$$V(s, x, T) = \mathrm{E}^{s,x}[v(s, T, X, u^*, c^*)] \tag{10}$$

where $\mathrm{E}^{s,x}[\cdot]$ is the expectation conditioned on $X_s = x$. According to the dynamic programming principle for stochastic control problems [10], the HJBE governing spatio-temporal evolution of the value function $V = V(s, x, T)$ is expressed as

$$\sup_{u \in U, c \in C} \left\{ A^{u,c}V - f(u) + \chi_{\{s \geq T_c\}}cx \right\} = A^{u^*,c^*}V - f(u^*) + \chi_{\{s \geq T_c\}}c^*x = 0. \tag{11}$$

The optimal controls $u = u^*$ and $c = c^*$ are analytically expressed through the value function V as

$$u^* = \beta^{-1}\min\{\max\{0, \ln(k_0\alpha\gamma) - \ln(\beta\omega)\}, 1\} \tag{12}$$

and

$$c^* = \chi_{\{\gamma \leq 1\}}c_\mathrm{M}, \tag{13}$$

respectively where the notation $\gamma = x\frac{\partial V}{\partial x}$ has been used. In this model, c^* is thus given as a bang-bang type control for each (s, x). The computational domain of the HJBE (11) is set as $\Omega = (0, L)$ with a positive constant L (kg) specified later. The HJBE (11) has to be equipped with appropriate terminal and boundary conditions for its well-posedness. The terminal condition to be equipped with the HJBE is set as $V_{s=T} = 0$ and the boundary conditions as $V_{x=0} = 0$ and $\frac{\partial V}{\partial x}\big|_{x=L} = 0$, respectively. The former boundary condition means that when the population of *P. altivelis* becomes extinct in the river system, the total profit V is 0, and the latter means that the total profit V doesn't increase when the population of *P. altivelis* exceeds L. Solving the HJBE yields the optimal controls u^* and c^* in the spatio-temporal domain $(0, T) \times (0, L)$.

2.3 Indices for Validating Optimal Management Strategy

Once the optimal controls u^* and c^* are computed from the HJBE, a variety of indices for validating optimal management strategy can be obtained considering the link between SDEs and terminal and boundary value problems of linear PDEs [10]. The index J defined as

$$J = \mathrm{E}^{s,x}\left[\int_s^T g(X_t)\mathrm{d}t\right] \tag{14}$$

with an univariate function g solves the Kolmogorov's backward equation (KBE)

$$A^{u^*,c^*}J + g = 0 \tag{15}$$

subject to appropriate terminal and boundary conditions, which are $J = 0$ at the time $t = T$, $J_{x=0}$ and $\frac{\partial J}{\partial x}\big|_{x=L} = 0$. Hereafter, the notation $J_0 = J(0, X_0)$ is used for the sake of brevity. Ecologically and economically relevant $g(X_t)$ can be $k(u_t^*)X_t$, $\chi_{\{t \geq T_c\}} c_t^* X_t$, and, $f(u_t^*)$ with which the index J_0 represents the total weight of predated $P.$ $altivelis$ by $P.$ $carbo$, the total weight of harvested $P.$ $altivelis$ as fish catches, and the total cost to prevent the feeding damage from $P.$ $carbo$, respectively. The indices with the above-mentioned three $g(X_t)$ are denoted in order as $J_{0,\mathrm{pr}}$, $J_{0,\mathrm{cau}}$, and $J_{0,\mathrm{cos}}$, respectively.

3 Application

3.1 Numerical Scheme

The Conforming Petrov-Galerkin Finite Element (CPGFE) scheme is used for approximating solutions to the HJBE (11) and the governing equations of the indices (15). The CPGFE scheme is based on the fitting technique where both the trial and test functions are determined from exponential analytical solutions to element-wise local two-point boundary value problems [13]. Spatial and temporal discretization procedure of the scheme is provided in Yoshioka et al. [13] and is therefore not presented here. For steady linear problems, numerical solutions with the scheme applying an appropriate lumping method to the mass matrix guarantee the parabolic discrete maximum principle [14]. The scheme has been applied to numerically solving a nonlinear elliptic problem for upstream fish migration in open channel flows [15]. Computational accuracy of the scheme for linear time-independent problems has been verified, and theoretical and numerical analyses indicated that its accuracy in space is at least first order in the L^∞-sense at each node. In this paper, a Picard iteration method is equipped in the spatial discretization procedure of the CPGFE scheme so that nonlinearity of the HJBE can be numerically handled at each time step.

3.2 Computational Conditions

Parameters of the present mathematical model are estimated from the collected data in and around Hii River system, San-in area, Japan where most part of population of $P.$ $altivelis$ are introduced through intensive release events under the initiative of Hii River fishery cooperatives. The total length of the mainstream and the catchment area of the river system are 153 (km) and 2,070 (km^2), respectively [16]. Hydrological characteristics of Hii River are well documented and statistically analyzed in detail in Sato et al. [17, 18], and are not explained in this paper. Two brackish water bodies, which are Lake Shinji and Lake Nakaumi from upstream, are connected to the main-stream of Hii River. Hatched larval fishes of *P. altivelis* are thought to descend to Lake Shinji; however, it is not known where actually the larval fishes survive during winter. Constituent members and union members of Hii River cooperatives say that the number of juvenile fishes ascending the river during spring is significantly decreasing [8]. In this river system, juvenile fishes of *P. altivelis* are released in its midstream and upstream reach during May. The fishes grow up in the system and spawn in down-stream reaches during October to November.

The initial time $t = 0$ for numerical computation is set on a day in May. The terminal time T is set as 180 (day), which is in the autumn in the same year. The parameters r and K are set as 8.7×10^{-2} (1/day) and $20X_0$ (kg), respectively, con-sidering a deterministic counterpart of (1) with $k = c = 0$ [19]. The natural mortality R is set as 4.6×10^{-3} (1/day) [20]. The range of the parameter σ is assumed to be restricted to $\sigma < 0.4$ (1/day$^{1/2}$), which is the condition that (1) with $k = c = 0$ has a non-trivial and non-singular PDF for sufficiently large T [21]. The value of $\sigma = 0.25$ (1/day$^{1/2}$) is thus assumed. The other parameters are $k_0 = 4.0 \times 10^{-3}$ (1/day) [8], $c_M = 0.01$ (1/day), and $\omega = 1$ (kg/day). $\alpha = 1$ (−) is fixed unless otherwise specified. The initial condition is $X_0 = 1.5 \times 10^3$ (kg), which amounts to 3.0×10^5 juveniles of *P. altivelis* [8]. The domain of the population x is set as $\Omega = (0, 6.0 \times 10^4)$ (kg), which is discretized into a mesh with 300 elements and 301 nodes. The time increment for temporally integrating the HJBE is 0.01 (day). Increasing spatial and temporal reso-lution of the computation does not significantly change the obtained computational results below.

3.3 Computational Results

Figures 1, 2 and 3 plot the computed value function V, the optimal control u^*, and the optimal control c^*, respectively. The panels (a)–(c) in each figure present the compu-tational results with $\beta = 1$, $\beta = 5$, and $\beta = 10$, respectively. The solution V for each β does not have numerical oscillations, indicating that the present CPGFE scheme can reasonably handle the HJBE derived in this paper. Figures 2(a)–(c) indicate that the optimal control u^* significantly depends on the inefficiency β of the countermeasure, and the optimal strategy common to the cases with small and moderate β is to intensively

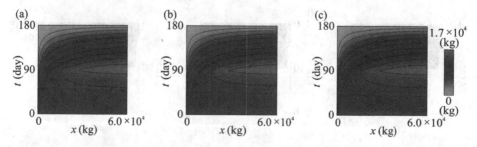

Fig. 1. Value function V for (a) $\beta = 1$, (b) $\beta = 5$, and (c) $\beta = 10$.

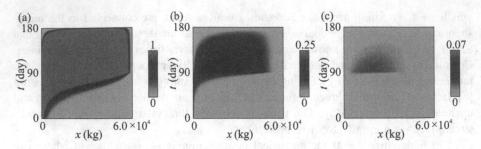

Fig. 2. Optimal control u^* for (a) $\beta = 1$, (b) $\beta = 5$, and (c) $\beta = 10$.

prevent feeding damage by *P. carbo* after the opening time T_c except for the neighborhood of the terminal time T. In these cases, it is also indicated that the intensive countermeasure should also be performed before the opening time T_c if the population of *P. altivelis* is small. For the largest β, the optimal strategy is not to perform the countermeasure. On the other hand, Figs. 3(a)–(c) indicate that the optimal control c^* does not significantly depend on β; the optimal strategy is to harvest *P. altivelis* after the opening time T_c except when the population of *P. altivelis* is small.

Comparisons of a variety of countermeasures to prevent the feeding damage, which are characterized through the non-dimensional parameters α and β, are next carried out. The indices considered are the total weight of predated *P. altivelis* ($J_{0,pr}$), the total weight of caught *P. altivelis* ($J_{0,cau}$), and the total cost to prevent the feeding damage by *P. carbo* ($J_{0,cos}$) presented in Sect. 2.3. The values examined are $\alpha = 0.1i$ and $\beta = j$ where i and j are nonnegative integers such that $0 \le i, j \le 10$. In total 121 numerical computations are performed for each index. Figures 4(a)–(c) plot the computed indices $J_{0,pr}$, $J_{0,cau}$, and $J_{0,cos}$ in the $\alpha - \beta$ phase space. These figures can validate the countermeasures in terms of the indices. The focus here is comparing the methods with high efficiency and low effectiveness (small β and small α) and those with low efficiency and high effectiveness (large β and large α). This is because countermeasures with low efficiency and low effectiveness (large β and small α) should not be used in practice, and those with high efficiency and high effectiveness (small β and large α) would merely exist; all the existing countermeasures would have both advantages and disadvantages on efficiency and effectiveness. Figures 4(a) and (c) indicate that the

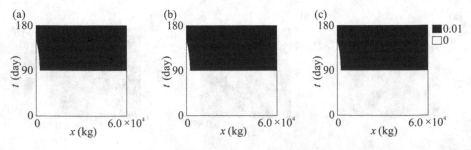

Fig. 3. Optimal control c^* for (a) $\beta = 1$, (b) $\beta = 5$, and (c) $\beta = 10$.

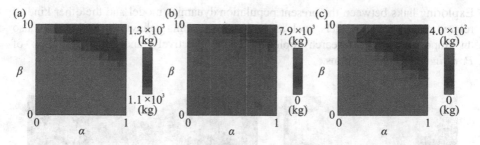

Fig. 4. The indices $J_{0,\text{pr}}$, $J_{0,\text{cau}}$, and $J_{0,\text{cos}}$ for a variety of countermeasures.

method with higher efficiency and lower effectiveness (small β and small α) can more effectively reduce the total weight of predated *P. altivelis* and the cost to preventing the feeding damage. In addition, Fig. 4(b) indicates that the method with higher efficiency and lower effectiveness more effectively increases the total fish catch. The obtained computational results thus recommend using a countermeasure with higher efficiency and lower effectiveness for managing the population of *P. altivelis*.

4 Conclusions

This paper applied a stochastic control theory to optimal management problem of population dynamics of *P. altivelis* in river systems subject to predation by *P. carbo* and fishing activity by human. Application of the dynamic programming principle to the SDE of population dynamics led to the HJBE governing the optimal strategy to maximize the sum of the cost to prevent the feeding damage and the profit of harvesting the fish. The model was applied to a real problem of inland fishery management in Hii River, Japan where most of the juvenile fishes are thought to be introduced through release events in spring. The CPGFE scheme successfully discretized the HJBE and computed the value function and the optimal management strategies under a variety of countermeasures. The presented mathematical and numerical modelling can serve as an effective tool to control population dynamics of *P. altivelis* in river systems in Japan.

Future research will extend the present mathematical model to a model with multiple habitats as considered in Unami et al. [12]. Yoshii Weir and Hinobori Weir installed in the mainstream of Hii River (Photo 2) have been thought to be physical barriers for upstream migration of juvenile *P. altivelis* that possibly fragment habitats of the fish. Another interesting extension of the present model is to consider hydraulic and hydrological conditions into the coefficients and parameters, such as the relations between water quality indices and the growth rate [22, 23]. These extensions would reduce to numerically solving an HJBE with larger number of independent variables and/or higher nonlinearity. Mathematical analysis on the HJBE of the present and extended models will also be performed for better comprehending their mathematical properties. Possible difficulties to be encountered in the mathematical analysis include degeneration of the diffusion term [24, 25] and discontinuity of the optimal controls. Such solutions should be dealt with from the viewpoint of viscosity solutions [26, 27].

Exploring links between the present population dynamics model and the other kind of related mathematical models, such as the local swimming behavior model [15], is also an important future research topic for more effectively analyzing migration of *P. altivelis* in river systems.

Photo. 2. Yoshii Weir (left panel) and Hinobori Weir (right panel) installed in the mainstream of Hii River.

Acknowledgements. We thank Dr. Kayoko Kameda of Lake Biwa Museum, and officers in Hii-River Fishery Cooperatives, Yonago Waterbirds Sanctuary, and the Ministry of Environment for their valuable comments and providing data. This research is supported by the River Fund in charge of The River Foundation and JSPS Research Grant No.15H06417.

References

1. Takahashi, I., Azuma, K.: The Up-to-now Knowledge Book of Ayu. Tsukiji-shokan, Tokyo (2006). (in Japanese)
2. MAFF: Statistics of catch and production amounts of aquatic resources in Japan during 2015 (2015). http://www.maff.go.jp/j/tokei/sokuhou/gyogyou_seisan_13/ (in Japanese)
3. Tanaka, Y., Iguchi, K., Yoshimura, J., Nakagiri, N., Tainaka, K.: Historical effect in the territoriality of ayu fish. J. Theor. Biol. **268**(1), 98–104 (2011)
4. Iguchi, K., Tsuboi, J., Tsuruta, T., Kiryu, T.: Foraging habits of Great Cormorant in relation to released Ayu stocks as a food source. Aquacult. Sci. **56**(3), 415–422 (2008)
5. Katano, O., Abe, S., Nakamura, T.: Relationships between ayu Plecoglossus altivelis and other organisms in stream communities. Bull. Fish. Res. Agen. Suppl. **5**, 203–208 (2006)
6. Yamamoto, M.: What Kind of Bird is the Great Cormorant. http://www.naisuimen.or.jp/jigyou/kawau/01-1.pdf (in Japanese)
7. Yamamoto, M.: Stand face to face with Great Cormorant II. http://www.naisuimen.or.jp/jigyou/kawau/03-1.pdf (in Japanese)
8. Hii River Fishery Cooperative. Personal Communication (2015)
9. Tsoularis, A., Wallace, J.: Analysis of logistic growth models. Math. Biosci. **179**(1), 21–55 (2002)
10. Øksendal, B.: Stochastic Differential Equations. Springer, Berlin (2003)
11. Doyen, L., Thébaud, O., Béné, C., Martinet, V., Gourguet, S., Bertignac, M., Fifas, F., Blanchard, F.: A stochastic viability approach to ecosystem-based fisheries management. Ecol. Econ. **75**, 32–42 (2012)

12. Unami, K., Yangyuoru, M., Alam, A.H.M.B.: Rationalization of building micro-dams equipped with fish passages in West African savannas. Stoch. Environ. Res. Risk Assess. **26**(1), 115–126 (2012)
13. Yoshioka, H., Unami, K., Fujihara, M.: Mathematical analysis on a conforming finite element scheme for advection-dispersion-decay equations on connected graphs. J. JSCE Ser. A2 **70**(2), I_265–I_274 (2014)
14. Mincsovics, M.E.: Discrete and continuous maximum principles for parabolic and elliptic operators. J. Appl. Math. Comput. **235**(2), 470–477 (2010)
15. Yoshioka, H., Unami, K., Fujihara, M.: A Petrov-Galerkin finite element scheme for 1-D time-independent Hamilton-Jacobi-Bellman equations. J. JSCE Ser. A2. **71**, I_149–I_160 (2016). (in press)
16. MLIT (2002). http://www.mlit.go.jp/river/toukei_chousa/kasen/jiten/nihon_kawa/87072/87072-1 (in Japanese)
17. Sato, H., Takeda, I., Somura, H.: Secular changes of statistical hydrologic data in Hii river basin. J. JSCE Ser. B1 **68**(4), 1387–1392 (2012). (in Japanese with English Abstract)
18. Sato, H., Takeda, I., Somura, H.: Secular changes of statistical dry-season discharges in Hii river basin. J. Rainwater Catchment Syst. **19**(2), 51–55 (2014). (in Japanese with English Abstract)
19. Miyaji, D., Kawanabe, H., Mizuno, N.: Encyclopedia of Freshwater Fishes in Japan. Hoiku-sha, Tokyo (1963). (in Japanese)
20. Murayama, T.: Why population of Ayu annually changes? In: Furukawa, A., Takahashi, I. (eds.) River Works to Grow up Ayu. Tsukiji-shokan, Tokyo (2010). (in Japanese)
21. Grigoriu, M.: Stochastic Calculus: Applications in Science and Engineering. Birkhäuser, Boston (2002)
22. Breitburg, D.L., Rose, K.A., Cowan, J.H.: Linking water quality to larval survival: predation mortality of fish larvae in an oxygen-stratified water column. Mar. Ecol. Prog. Ser. **178**, 39–54 (1999)
23. Budy, P., Baker, M., Dahle, S.K.: Predicting fish growth potential and identifying water quality constraints: a spatially-explicit bioenergetics approach. Environ. Manage. **48**(4), 691–709 (2011)
24. Chernogorova, T., Valkov, R.: Analysis of a finite volume element method for a degenerate parabolic equation in the zero-coupon bond pricing. Comput. Appl. Math. **34**(2), 619–646 (2015)
25. Chernogorova, T., Valkov, R.: Finite volume difference scheme for a degenerate parabolic equation in the zero-coupon bond pricing. Math. Comput. Model. **54**(11), 2659–2671 (2011)
26. Crandall, M.G., Lions, P.L.: Viscosity solutions of Hamilton-Jacobi equations. Trans. Am. Math. Soc. **277**(1), 1–42 (1983)
27. Fleming, W.H., Soner, H.M.: Controlled Markov Processes and Viscosity Solutions. Springer Science & Business Media, New York (2006)

Fault Displacement Simulation Analysis of the Kamishiro Fault Earthquake in Nagano Prefecture Using the Parallel Finite Element Method

Yuta Mitsuhashi[1](✉), Gaku Hashimoto[2], Hiroshi Okuda[2], and Fujio Uchiyama[1]

[1] Disaster Reduction and Environmental Engineering Deptartment,
Kozo Keikaku Engineering, Inc., 4-5-3 Chuo, Nakano-Ku, Tokyo 164-0011, Japan
yuta-mitsuhashi@kke.co.jp
[2] Graduate School of Frontier Sciences, University of Tokyo, 5-1-5 Kashiwanoha,
Kashiwa-shi, Chiba 277-8563, Japan

Abstract. In recent years, the evaluation of fault displacement has been required for evaluating the soundness of underground structures during an earthquake. Fault displacement occurs as the result of the rupture of the earthquake source fault, and studies have been conducted using the finite difference method, the finite element method, etc. The present study used the nonlinear finite element method to perform a dynamic rupture simulation analysis of the Kamishiro fault earthquake in Nagano Prefecture on November 22, 2014. A model was prepared using a solid element for the crust and a joint element for the fault surface. The Kamishiro fault earthquake in Nagano was a reverse-fault earthquake whose fault plane included a part of the Kamishiro fault and extended northward from there. The total extent was 9 km, and the surface fault displacement confirmed was approximately 1 m at maximum. Initial stress was applied to the fault to intentionally rupture the hypocenter to perform a propagation analysis of the rupture, and the displacement and response time history obtained in the analysis were compared with observational records. At this time, joint elements according to Goodman et al. that had been expanded were introduced to the finite element method code FrontISTR, which can analyze large-scale models, and the simulation analysis was performed.

Keywords: Fault distance · FEM · Joint element · Dynamic rupture simulation · Parallel computing

1 Introduction

In recent years, the evaluation of fault displacement has been required for evaluating the soundness of underground structures during an earthquake. Fault displacement occurs as the result of the rupture of the earthquake source fault, and studies have been conducted using the finite difference method (i.e. [1]) and the finite element method (i.e. [2]). In particular, many studies centering on the finite difference method have been performed using dynamic rupture simulation analysis, which reproduces the spontaneous rupture process of a fault using a slip-weakening model (i.e. [1]).

© Springer Science+Business Media Singapore 2016
S.Y. Ohn and S.D. Chi (Eds.): AsiaSim 2015, CCIS 603, pp. 102–109, 2016.
DOI: 10.1007/978-981-10-2158-9_9

The present study used the 3D nonlinear finite element method to perform a simulation analysis of the Kamishiro fault earthquake in Nagano Prefecture on November 22, 2014. The Kamishiro fault earthquake in Nagano (Mj 6.7, *Mw* 6.2) was a reverse-fault earthquake whose fault plane included a part of the Kamishiro fault and extended northward from there, with a length of approximately 20 km and a depth of approximately 10 km. The total extent was 9 km, and the surface fault displacement confirmed was approximately 1 m at maximum [3]. Also, a maximum acceleration of approximately 600 cm/s^2 was observed in the K-NET [4] observation points near the fault. A 40 km × 40 km × 20 km model that included the earthquake source fault was prepared using a solid element for the crust and a joint element for the fault. A model was created for the rupture process of the fault according to the nonlinear constitutive law incorporating stress drop, and for the initial conditions, initial stress was applied to the joint element to rupture the fault, and the ground surface response time history caused by propagation of the rupture was compared with the observational records of K-NET. Also, the surface fault displacement observed in studies of the actual site after the earthquake was compared with the displacement obtained in the present analysis. This research was conducted using the finite element method code FrontISTR [5], which can perform parallel computation of large-scale models. This allowed a wide area of the crust to be analyzed using relatively fine mesh.

2 Analytical Conditions of the 3D Finite Element Method

2.1 Creating a Model of the Fault by Joint Elements

In the present study, a model of the fault was created by the joint elements shown in Fig. 1. Those joint elements are finite elements that easily simulate the contact/sliding/exfoliation between two physical bodies. Several joint elements have been proposed, such as those proposed by Goodman et al. [6] and those that have been formulated on the basis of 3D isoparametric elements. The shape of the elements according to Goodman et al. was hypothesized to be rectangular, and the deformation between the two contacting surfaces is represented by the 6-mode combination shown in Fig. 2. The authors expanded this to the desired shape of a triangle or quadrilateral, and implemented FrontISTR. This allowed a high-precision analysis to be performed, even with distorted mesh.

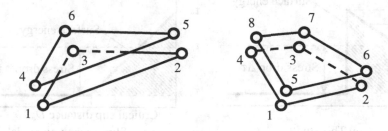

Fig. 1. Joint elements

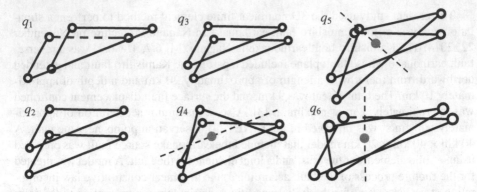

Fig. 2. Deformation mode of the joint elements

The relationship between the shearing stress and relative displacement of the joint element used in this study is shown in Fig. 3(a). The joint elements have a bi-directional degree of freedom with respect to the translation deformation in the surface (q_1 and q_2 shown in Fig. 2), so the shearing stress τ and the relative displacement ε are evaluated in Eqs. (1) and (2). However, f_{q1} and f_{q2} are the shearing stress according to modes q_1 and q_2, and δ_{q1} and δ_{q2} are the displacement of modes q_1 and q_2.

$$\tau = \sqrt{f_{q_1}^2 + f_{q_2}^2} \tag{1}$$

$$\delta = \sqrt{\delta_{q_1}^2 + \delta_{q_2}^2} \tag{2}$$

As shown in Fig. 3(a), sliding rupture occurs when the shearing stress τ of joint elements reaches the yield stress τ_y, and a stress drop to τ_0 occurs. According to past studies (i.e. [1]), the behavior of the shearing stress after sliding rupture is that it does not rapidly drop to τ_0 but has a certain degree of inclination, and with the slip-weakening model in the finite difference method, a frequently used model is one in which the shearing stress drops linearly to critical slip displacement D_c shown in Fig. 3(b). For convenience sake,

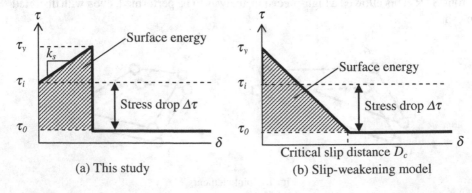

(a) This study

(b) Slip-weakening model

Fig. 3. Stress displacement relationship

the present study used a model in which the shearing stress dropped to τ_0 the instant that sliding rupture occurred. Also, according to past studies [2], the analytical results depend on the relative quantity $\Delta\tau = \tau_y - \tau_0$, not on the absolute quantity τ_0. Therefore, this study used an analysis that focused on the relative value $\Delta\tau$ when $\tau_0 = 0$MPa. In addition, referring to Dan et al. ([1]), the initial shearing stiffness k_s of the joint elements was specified so that the surface energy calculated from $D_c = 25$ cm in the slip-weakening model was equivalent, and the vertical stiffness k_v was made to be linear and a sufficiently rigid value.

2.2 Movement Equation

The movement equation is given by Eq. (3).

$$M\ddot{x} + C\dot{x} + Kx = F \tag{3}$$

Here, x is the displacement; M, C, and K are the mass, damping, and stiffness matrix, respectively; F is the external force vector. For damping, stiffness-proportional damping was used, and referring to the maximum transmission frequency of the model and the study by Mizumoto et al. [2], it was established that there would be 2 % damping at 1 Hz. Also, the effect of the reflected wave was eliminated by setting the viscous boundary to the model periphery that excluded the crust surface. According to the constitutive law of the joint elements shown in Fig. 3(a), this problem had non-linearity, so a convergent calculation according to the Newton-Raphson method was performed.

3   Analysis Model

The model used in analysis is shown in Fig. 4, and the fault parameters are shown in Table 1. The fault shape was established by referring to the AIST active fault database [7] and aftershock distribution, and a strike angle of 12°, dip angle of 50°, length of 18 km, and depth of 10 km (width of 12.2 km) were established. As for generally used physical values, $\nu = 0.25$ was hypothesized with the share modulus of stiffness μ set to 30 GPa and the unit weight γ was set to 2.5 t/m3. Referring to the records on the hypocenter of this earthquake, the hypocenter was set to a position at a 5-km depth, and the τ_y of the joint elements was set at a large value. The Kamishiro fault earthquake was a reverse-fault earthquake. But the earthquake was also indicated to have been accompanied by a left-lateral slip. So we assume the direction of initial stress λ as 90° (pure reverse-fault), 75°, 60° and 45°, and we compare their results(See Fig. 4). By setting an initial stress, rupture would occur at the same time that the analysis started. Since ground surface displacement was not observed on the north side of the fault, the shape of the fault surface might not have been a simple rectangle, but for convenience sake, this study used a rectangular fault.

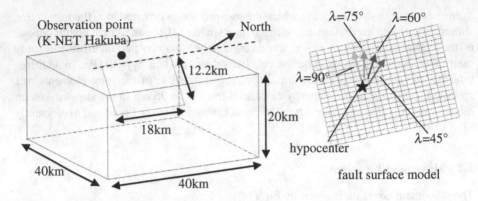

Fig. 4. Analysis model.

Table 1. Fault parameters

Fault width	W	12.2	km
Fault length	L	18.0	km
Strike angle	θ	12	degree
Dip angle	δ	50	degree
Stress drop	$\Delta\tau$	1.00	MPa
Fault shearing stiffness	k_s	1.20×10^4	kN/m/m^2
Fault vertical stiffness	k_v	1.20×10^4	kN/m/m^2

The stress drop was uniformly set to 1 MPa, the condition in which the moment magnitude Mw, calculated from the fault displacement Δu of each element in the final state of the analysis by using Eqs. (4) and (5), with the fault surface as Σ, was close to the value observed in the actual earthquake. τ_y was established according to Eq. (6), referring to Andrew [8].

$$M_0 = \int_{\Sigma} \mu \Delta u dS \tag{4}$$

$$Mw = \frac{2}{3} \log_{10} M_0 - 6.1 \tag{5}$$

$$\tau_y = 1.6 \Delta\tau \tag{6}$$

The maximum frequency which can be covered by the mesh f_{max}, that is calculated in Eq. (7), is 0.875 Hz when $n = 4$. However, β is the shear wave velocity at the crust.

$$f_{max} = \frac{\beta}{n\Delta L} \tag{7}$$

The analysis was a consecutive nonlinear response analysis according to the Newmark-β method (parameter $\beta = 0.25$, $\gamma = 0.5$). The integration time of the analysis Δt was 0.01 s, and the duration T was 20 s.

4   Analytical Results

The analytical results are shown. The simulation analysis matched well when initial stress direction λ is 60°. So deformation diagram, fault rupture time, fault displacement only the case is shown as a representative case.

4.1   Model Deformation Diagram

The deformation diagram of the analysis final time is shown in Fig. 5.

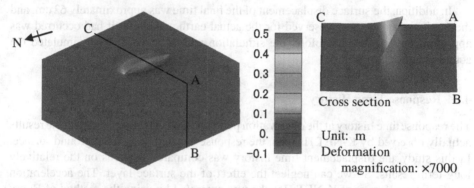

Fig. 5.   Analysis model deformation diagram and vertical displacement contour diagram.

4.2   Fault Rupture Time and Fault Displacement

A contour diagram of the time that each joint element of the fault surface ruptured is shown in Fig. 6(a). The average rupture propagation time V_r was evaluated as 3 km/s from Fig. 6(a), and the S-wave speed β at the crust was 3.46 (km/s); therefore, the ratio V_r/β equals 0.87. According to recipe [9], the general ratio V_r/β is 0.72, but the fact that larger values have been obtained is also described, so the results of this test are not believed to greatly diverge from the reality. Also, the rupture propagation speed has been confirmed to increase as a result of changing the surface energy required at the time of rupture (See Fig. 3).

Contour diagrams of the fault displacement that occurred in the joint elements in the analysis final time are shown in Fig. 6(b). The seismic moment M_0, which is the index of the size of the earthquake, computed from the fault displacement of the final time was 2.77×10^{18} Nm, and the moment magnitude Mw calculated from M_0 was 6.2. This agreed well with the results for the actual earthquake according to the CMT analysis presented by the Japan Meteorological Agency in which the seismic moment was 2.98×10^{18} Nm and the moment magnitude was 6.2.

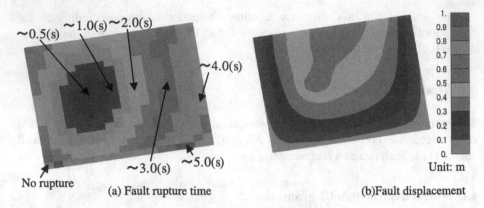

(a) Fault rupture time   (b)Fault displacement

Fig. 6. Fault rupture time and fault displacement.

In addition, the surface displacement of the final time was approximately 62 cm, and the vertical displacement observed for the actual earthquake after it had occurred was approximately 90 cm. Therefore, the simulation was believed to have simulated the actual phenomenon well.

4.3 Response Time History

The response time history at the observation point location was compared with the results actually observed by K-NET Hakuba, the response acceleration on the ground surface. In this study, the displacement time history was compared to focus on the relatively long-term results. So we can neglect the effect of the surface layer. The acceleration response time history of K-NET Hakuba was corrected by using the method of Boore et al. [10], and the displacement time history waveform that were obtained by time integration were compared with the waveform obtained in the analysis. A comparison of the displacement time history obtained from the observational records with the

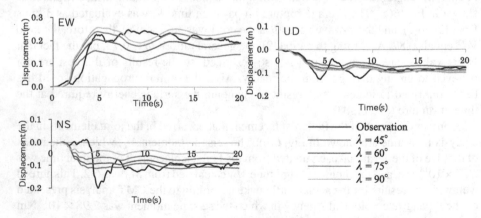

Fig. 7. Comparison of displacement response time history. (Color figure online)

analytical results is shown in Fig. 7. However, the time axis origin has been adjusted as appropriate.

Because the observation points of K-NET Hakuba were relatively close to the fault, a relatively large, permanent displacement occurred in the analysis final time. The simulation analysis matched well when initial stress direction is 60°.

5   Conclusion

A fault dynamic simulation analysis of the Kamishiro fault earthquake in Nagano Prefecture was performed using the finite element method. By adjusting the stress drop $\Delta\tau$, analytical results close to the seismic moment in the actual earthquake were able to be obtained, and the results for the surface fault displacement and displacement time history matched well with the observations of the actual earthquake. In the future, a more comprehensive study that includes a more detailed fault shape and physical property value variations is desired.

Acknowledgments.   This study used the K-NET strong motion seismograms from the National Research Institute for Earth Science and Disaster Prevention.

References

1. Dan, K., Muto, M., Torita, H., Ohhashi, Y., Kase, Y.: Basic examination on consecutive fault rupturing by dynamic rupture simulation. In: Annual Report on Active Fault and Paleoearthquake Researches, no.7, pp. 259–271 (2007). (In Japanese)
2. Mizumoto, G., Tsuboi, T., Miura, F.: Fundamental study on fault rupture process and earthquake motions on and near a fault by 3D-FEM. J. JSCE, no.780, I-70, 27–40 (2005). 1 (In Japanese)
3. Japan Association for Earthquake Engineering: Report about the Earthquake at the Nagano Prefecture north in 2014 (2015). (In Japanese)
4. National Research Institute for Earth Science and Disaster Prevention: Strong-motion Seismograph Networks. http://www.kyoshin.bosai.go.jp/. Accessed 01 Aug 2015
5. FrontISTR Workshop HP. http://www.multi.k.u-tokyo.ac.jp/FrontISTR/. Accessed 01 Aug 2015 (In Japanese)
6. Goodman, R.E.: Methods of geological engineering in discontinuous rocks, Chap. 8, pp. 300–368. West Publishing Company (1976)
7. National Institute of Advanced Industrial Science and Technology: Active fault data base. https://gbank.gsj.jp/activefault/index_gmap.html. Accessed 01 Aug 2015 (In Japanese)
8. Andrews, D.J.: Rupture velocity of plane strain shear cracks. J. Geophys. Res. **81**, 5679–5687 (1976)
9. Earthquake Research Committee, Headquarters for Earthquake Research Promotion, Strong motion forecasting method of the earthquake which specified an earthquake source fault (Recipe). Revision on 21 Dec. 2009 (In Japanese)
10. Boore, D.M., Stephens, C.D., Joyner, W.B.: Comments on baseline correction of digital strong-motion data: examples from the 1999 Hector Mine, California, Earthquake. Bull. Seismol. Soc. Am. **92**(4), 1543–1560 (2002)

Prediction of 137Cs-Contaminated Soil Losses by Erosion Using USLE in the Abukuma River Basin, Japan

Carine J. Yi[✉]

International Research Institute of Disaster Science, Tohoku University, Sendai, Japan
carineyi@irides.tohoku.ac.jp

Abstract. The 2011 Great East Japan Earthquake and tsunami triggered a significant nuclear power plant accident. Subsequent measurements of the concentration of cesium-137 (137Cs) showed that the litter and surface layers in the forest areas near the plant were significantly contaminated. This study applied the Universal Soil Loss Equation (USLE), which has been widely used to estimate soil losses from erosion, in the Abukuma River Basin. The greatest soil loss was predicted to be 1762.75 t • yr^{-1} • ha^{-1}. To predict losses of 137Cs-contaminated soil, a 137Cs-soil transfer factor was applied in place of a crop factor, and it yielded an average contaminated-soil loss rate of 190.65 t • yr^{-1} • ha^{-1}, whereas the standard USLE calculation yielded an estimated average soil loss rate of 184.14 t • yr^{-1} • ha^{-1}. Higher soil losses were predicted in steeper areas west of the river. However, contaminated soil may be deposited along a comparatively flat area, such as that on the east side of the river.

Keywords: Fukushima Daiichi Nuclear Power Station explosion · Cesium-137 · USLE · Abukuma river basin

1 Introduction

Studies of nuclear events were first conducted during World War II and focused on topics such as the life cycle, mechanisms, and environmental impacts of fallout. Cesium-137 (137Cs) is among the fission products that generate the most concern because of its beta and gamma emissions, high radioactivity, and relatively long, 30-year half-life (Okumura 2003), and it is only produced during nuclear fission events (Ritchie and McHenry 1990). According to Carter and Moghissi (1977), 137Cs was first released during early nuclear tests in 1945, and studies on the global dispersal of 137Cs into the environment began with high-yield thermonuclear tests in November 1952. Perkins and Thomas (1980) and Ritchie and McHenry (1990) developed the earliest studies of the dispersal of 137Cs. 137Cs fallout is strongly influenced by local precipitation patterns and rates (Longmore 1982). However, 137Cs concentrates on the soil surface when the soil is left undisturbed (Gerspar 1970; Ritchie et al. 1972). Once in contact with the soil, 137Cs is firmly adsorbed to finer soil particles, such as clay, and further movement of 137Cs by natural chemical processes is limited (Tamura, 1964; Ritchie et al. 1982; Guimarães et al. 2003). Rogowski and Tamura (1965) investigated the difference between predicted and experimental soil losses by spraying 137Cs onto areas of bare

© Springer Science+Business Media Singapore 2016
S.Y. Ohn and S.D. Chi (Eds.): AsiaSim 2015, CCIS 603, pp. 110–117, 2016.
DOI: 10.1007/978-981-10-2158-9_10

clipped meadow and tall meadow cover. Walton (1963) determined the vertical radio-activity profiles associated with weapons tests in several New Jersey soils by examining four radionuclides: strontium 90, ruthenium 106, 137Cs, and 144Cs. McHenry and Ritchie (1973) found that 137Cs concentrated in accumulation sites, such as floodplains, lakes, and reservoirs. Okumura (2003) described the life cycle of radioactive 137Cs and analyzed its material flow in the U.S. Guimarães et al. (2003) described the vertical distribution of 137Cs in the soil profile and showed its exponential decrease with increasing soil depth at undisturbed sites. Mizugaki et al. (2008) analyzed suspended sediment sources using 137Cs and 210Pb$_{ex}$ in Japan before the Fukushima Daiichi Nuclear Power Station explosion.

The 1986 Chernobyl accident, which was classified as level 7 (major accident) according to the International Nuclear Event Scale (Cardis et al. 2006), has been the only nuclear accident to provide in-field data that were used to identify contamination areas and trace radioactive materials, determine the impacts to human health from external or internal contamination via long-term monitoring, and evaluate the impact of radioactive fallout on the environment and agricultural activities. Subsequent to the Chernobyl accident, the Great East Japan Earthquake and Tsunami occurred in 2011 and triggered an explosion at the Fukushima Daiichi Nuclear Power Station. Residents within a 20-km radius of the Fukushima Daiichi Nuclear Power Station were forced to leave their homes. Japanese government agencies continue to measure the level of radioactive contamination in urban areas, agricultural areas, forests, rivers and ocean. A total of 3,793 observation locations are measured daily throughout the country and used to chart the current and daily average levels and other measurements of the radioactivity associated with this disaster (a national radioactivity information list is available online at http://new.atmc.jp/).

Since the Fukushima Daiichi Nuclear Power Station explosion in 2011, a number of measurement-based studies have been conducted. Yasunari et al. (2011) applied a Lagrangian particle dispersion model known as FLEXPART to estimate the total deposits of 137Cs by integrating daily observations of 137Cs in each prefecture in Japan. Tonosaki et al. (2012) measured radiation levels in surface soils and found that they contained greater concentrations of 134Cs and 137Cs relative to deeper soil layers. However, Hayashi et al. (2012) measured 134Cs and 137Cs contamination in areas around Mt. Tsukuba, Ibaraki, Japan and found that the levels were higher in forest than in open areas, such as paddy fields and rural towns. In addition, they found that the litter layers and surfaces in the forest were significantly contaminated.

Radioactive material released from a nuclear accident follows a process in which it falls to the ground surface, is partially retained in the soil and vegetation, and to some extent is then transported into rivers and distributed into bottom sediment (NIES, 2012). In this study, the Universal Soil Loss Equation (USLE) is used to estimate the transfer of 137Cs through soil erosion into the Abukuma River Basin, which is the closest river basin to the site of the incident. The USLE is an erosion model designed to predict the average rate of soil erosion for each feasible alternative combination of crop system and management practice in association with a specified soil type, rainfall pattern, and topography (Wischmeier and Smith 1978).

The Abukuma River runs from southwest to northeast in the Fukushima Prefecture, which has a population of 1,380,000. The river is 239 km long, and the total river network is 1,814 km long. The river basin area is 5,405 km^2. The river is a first-class fresh water resource that is surrounded by mountainous areas that have summit heights in excess of 1,000 m and occasionally experience heavy rains (MLIT; Water and Disaster Management Bureau 2003).

2 USLE Application

The USLE is a widely applied erosion model used to estimate soil loss on slopes due to runoff in specific field areas subjected to particular cropping and management systems (Warren et al. 2005). Wischmeier and Smith (1978) developed the USLE for conservation planners, and the soil equation is defined as follows:

$$A = RKLSCP \tag{1}$$

where
A = soil loss (tons/ha/year),
R = rainfall erosivity factor (tf • m^2/ha • h),
K = soil erodibility factor (h/m^2),
L = slope length factor,
S = slope steepness factor,
C = cover-management factor, and
P = supporting practice factor (management factor).

The variable A is the calculated soil loss per unit area. It is expressed in the same units selected for K and includes the same period selected for R. In practice, these parameters are usually selected such that A can be calculated in tons per hectare per year. However, other units may be selected if desired. The USLE is an erosion model designed to calculate long-term soil loss overages from sheet and rill under specified conditions. The model is also useful for construction sites and other non-agricultural conditions, although it does not predict deposition or calculate sediment yields from gully, stream bank, or streambed erosion. Other studies have applied the USLE to Japanese soil (Higa and Manmoto 2000; Imai and Ishiwatari 2006; Imai and Ishiwatari 2007; Unoki et al. 2010).

R is the rainfall-runoff erosivity factor, and it is calculated from daily precipitation data. In this study, the Automated Meteorological Data Acquisition System (AMeDAS) averages of September 2009 were reviewed to determine the R factor, which is defined as

$$R = \sum_{i=1}^{n} E_i I_{30i} \tag{2}$$

where I is the rainfall intensity (cm/h) and R is the amount of rainfall (cm). The variable E_i is defined as follows:

$$E_i = E_{0i} r_i. \tag{3}$$

K is the soil erodibility factor, and it represents the soil loss rate per erosion index unit for a specified soil type. The soil erodibility factor is measured on a unit plot defined as a 72.6-foot long, uniform 9 % continuous slope in clean-tilled fallow soil. The K factor according to soil type is summarized based on work by Imai and Ishiwatari (2007).

L is the slope length factor, and it is the ratio of soil loss from the field slope length to the soil loss from a 72.6-foot (22.13-meter) slope length under identical conditions.

S is the slope steepness factor, and it is the ratio of the soil loss from the field slope gradient to that from a 9 % slope under otherwise identical conditions. The S factor may be evaluated by combining the L factor for each land cell (Wischmeier and Smith 1978):

$$LS = (\lambda / 22.13)m \ (65.4 \sin2 \beta + 4.56 \sin \beta + 0.0654) \tag{4}$$

where m is 0.5, 0.4, 0.3, and 0.2 for $\tan \beta > 0.05$, $0.03 < \tan \beta \leq 0.05$, $0.01 < \tan \beta \leq 0.03$, and $\tan \beta \leq 0.01$, respectively (λ is the slope length in meter and ß is the slope angle in degrees).

C is the cover and management factor, and it is the ratio of the soil loss from an area with specific cover and management characteristics to that from an identical area under tilled, continuous fallow conditions. In Fukushima, the cover/management categories and corresponding C factors are as follows: orchard (0.2), farm (0.4), rice (0.01), livestock (0.01), flowers (0.01), and grass crops (0.05); these values were suggested by the Okinawa Science and Technology Promotion Center (OSTC) and are based on the Okinawa Prefecture land-use map (2001).

P is a supporting practice factor (management factor), and it takes on the following values: 0.3 for orchards, 0.3 for farms, 0.1 for rice, 0.1 for livestock, 0.1 for flowers, and 0.3 for grass crops.

3   Results

Figure 1 depicts the results of the USLE calculations. In the figure, the color changes progressively among adjacent regions within the range of white to black, with white indicating no soil loss and black indicating maximum soil loss. The standard USLE calculations (left panel) predicted an average loss of 184.14 t • yr^{-1} • ha^{-1} of 137Cs-contaminated soil from the Abukuma River Basin. The greatest predicted soil loss, 1762.75 t • yr^{-1} • ha^{-1}, was concentrated in the steep areas west and north of the center of the Abukuma River Basin.

After performing the standard USLE calculations, the P factor was replaced by the 137Cs transfer factor, Tp. The Radioactive Waste Management Center (1988) proposed 137Cs transfer implementation coefficients for agricultural products, for example, 0.00070 for tomatoes, 0.012 for beans, 0.0011 for radishes and 0.0010 for apples. The Tp value is calculated as follows:

$$Tp = \frac{137Cs \ concentration \ in \ agricultural \ products \ (Bq/kg)}{137Cs \ concentration \ in \ soil \ (Bq/kg)} \tag{5}$$

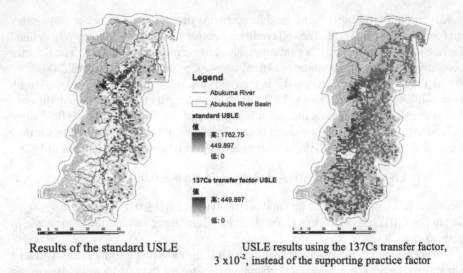

Results of the standard USLE   USLE results using the 137Cs transfer factor,
                                                3 x10⁻², instead of the supporting practice factor

Fig. 1. Results of USLE calculations

The right-hand panel in Fig. 1 depicts the distribution of eroded soil contaminated with 137Cs, which replaced the supporting practice factor as the soil transfer factor. A maximum soil loss of 449.897 t • yr⁻¹ • ha⁻¹ (an average of 190.65 t • yr⁻¹ • ha⁻¹) was predicted in the Abukuma River Basin using this calculation. Active soil loss was predicted upstream of the river, whereas other areas in the hydrology network indicated minimal or no soil loss.

The Abukuma River Basin is a mountainous region with 77,202 ha of cultivated area (Abukuma River Basin Risk Reduction Committee 2006), as shown in Fig. 2. In the study area, an average of 190.65 t • yr⁻¹ • ha⁻¹ (a maximum soil loss of 449.8s97 t • yr⁻¹ • ha⁻¹) of 137Cs-contaminated soil was predicted by the USLE to be lost through soil erosion.

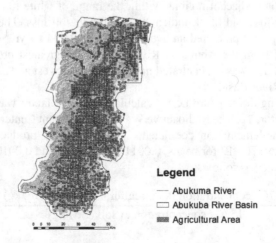

Fig. 2. Cultivated land in the Abukuma River Basin

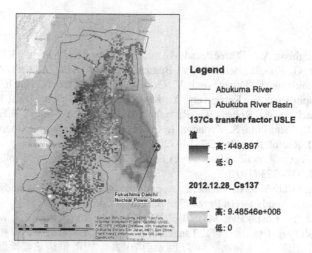

Fig. 3. 137Cs (2012.12.28) air concentrations overlaid on the results of the 137Cs transfer factor USLE

Higher soil erosion areas were predicted to occur in association with steep slopes and fallow soil; these areas were found west and north of the center of the study area. However, soil erosion occurs widely in the study area near the main streams of the Abukuma River. Figure 3 shows the 137Cs concentration data from December 28, 2012 overlaid with the results of the 137Cs transfer factor USLE for the 23 of 59 towns in Fukushima Prefecture that lie within the Abukuma River Basin. The predicted losses of contaminated soil on the east side of the river are small compared with those of the west side, as apparent in the figure from the numerous dark gray to black areas in the west, and the 137Cs concentrations on the east side of the river are still high. High 137Cs air concentrations gradually spread out from the east coast to the northwest and the southwest. Relatively lower 137Cs air concentrations occur along the Abukuma River, which consists largely of flatland areas.

4   Conclusions

137Cs has a 30-year half-life cycle and disappears naturally by radioactive decay. This study showed that 137Cs transfer can be estimated according to soil erosion rates in flatland areas that contain 137Cs contaminated soil and where predicted soil loss is less than in steep areas. Therefore, further attention might be required and the findings described here can help inform disaster recovery strategies. Intense rainfall may be a significant factor in the transportation of contaminated soil to downstream areas and should be monitored in future watershed risk management strategies.

References

Carter, M.W., Moghissi, A.A.: Three decades of nuclear testing. Health Phys. **33**(1), 55–71 (1977)

Cardis, E., Howe, G., Ron, E., Bebeshko, V., Bogdanova, T., Bouville, A., Carr, Z., Chumak, V., Davis, S., Demidchik, Y., Drozdovitch, V., Gentner, N., Gudzenko, N., Hatch, M., Ivanov, V., Jacob, P., Kapitonova, E., Kenigsberg, Y., Kesminiene, A., Kopecky, K.J., Kryuchkov, V., Loos, A., Pinchera, A., Reiners, C., Repachol, M., Shibata, Y., Shore, R.E., Thomas, G., Tirmarche, M., Yamashita, S., Zvonova, I.: Cancer consequences of the Chernobyl accident: 20 years on. J. Radiol. Prot. **26**, 127–140 (2006)

Gerspar, P.L.: Effect of American Beech trees on the gamma radioactivity in soils. Soil Sci. Soc. Am. Proc. **34**, 318–323 (1970)

Guimarães, M.F., Filho, V.F.N., Ritchie, J.: Application of cesium-137 in a study of soil erosion and deposition in Southeastern Brazil. Soil Sci. **168**(1), 45–53 (2003)

Hayashi, S (林誠二)., Koshikawa, M (越川昌美)., Watanabe, M (渡邊未来)., Watanabe, K (渡邊圭司)., Nishikori, T (錦織達啓)., Tanaka, A (田中敦).: 茨城県筑波山森林域からの放射性セシウム流出特性. 本陸水学会 77 回大会 2012 名古屋. 同講演要旨集 89 (2012). (in Japanese)

Higa, E (比嘉榮三郎)., Manmoto, H (満本裕彰).: USLE 式による土壌流出予測方法. 平成 12 年度流域赤土流出防止報告書(沖縄) (2000). (in Japanese)

Imai, K (今井啓)., Ishiwatari, T (石渡 輝夫).: Regional characteristics of USLE factors based on literate data. Extended abstract for 2006 the conference of Japanese society of Irrigation, drainage and rural engineering, 平成 18 年度農業土木学会大会講演会講演要旨集 (2006). (in Japanese)

Imai, K (今井啓)., Ishiwatari, T (石渡 輝夫).: 統計資料等を用いて整理した都道府県別の土壌侵食因の地域性について, 寒地土木研究所 月報 № 645 (July, 2007)

Ritchie, J.C., McHenry, J.R.: Application of radioactive fallout cesium-137 for measuring soil erosion and sediment accumulation rates and patterns: a review. Int. J. Environ. Qual. **19**(2), 215–233 (1990)

Ritchie, J.C., McHenry, J.R., Bubenzer, G.D.: Redistribution of fallout 137Cs in Brunner Creek watershed in Wisconsin. Wis. Acad. Sci. Arts Lett. **70**, 161–166 (1982)

Longmore, M.E.: The caesium-137 dating technique and associated applications in Australia - a review. In: Ambrose, W., Duerden, P. (eds.) Australian Archaeometry Conference; Sydney (Australia), 396 p., pp. 310–321. Australian National University, Canberra, 15-18 February 1982. ISBN 0 86784 239 3, 37 refs

Tonosaki, M (外崎真理雄)., Kaneko, S (金子真司)., Kiyono, Y (清野嘉之).: 放射性セシウムによる森林や木材への影響について. 木材情報 (日本木材総合情報センター, 森林総合研究所.) 249, 1–6 February 2012. (in Japanese)

McHenry, J.R., Ritchie, J.C.: Accumulation of fallout cesium 137 in soils and sediments in selected watersheds. Water Resour. Res. **9**(3), 676–686 (1973)

Ministry of Land, Infrastructure, Transportation and Tourism (MILIT)-Water and Disaster Management Bureau (国土交通省河川局): 武隈川水系の流域及び河川の概要(案)-参考資料 4-1, (Report for Abukuma River Basin management) 平成 15 年 11 月, November 2003. (in Japanese)

Okinawa Science and Technology Promotion Center (OSTC, former: (財) 亜熱帯総合研究所): 平成 13 年度 内閣府委託事業 「珊瑚礁に関する調査」: GIS 利用による陸域影響に関する調査研究 (2001). (in Japanese)

Perkins, R.W., Thomas, C.W.: Worldwide fallout. In: Hanson, W.C. (ed.), pp. 53–82. Technical Information Center, Oak Ridge (1980)

Ritchie, J.C., McHenry, J.R., Gill, A.C.: The distribution of Cs-137 in the litter and upper 10 centimeters of soil under different cover types in northern Mississippi: Health Physics, vol. 22, p. 197 (1972)

Rogowski, A.S., Tamura, T.: Movement of 137Cs by Runoff, Erosion and Infiltration on the Alluvial Captina Silt Loam. Health Phys. Soc. 11(12), December 1965

Warren, S.D., Mitasova, H., Hohmann, M.G., Landsberger, S., Iskander, F.Y., Ruzycki, T.S., Senseman, G.M.: Validation of a 3-D enhancement of the universal soil loss equation for prediction of soil erosion and sediment deposition. CATENA – Interdisc. J. Soil Sci. Hydrol. Geomorphol. Focusing Geoecology Landscape Evol. **64**, 281–296 (2005)

Okumura, T.: The material flow of radioactive cesium-137 in the U.S. 2000, United States Environmental Protection Agency (2003)

Yasunari, T.J. Stohl, A., Hayano, R.S., Burkhart, J.F., Eckjardt, S., Yasunari, T.: Cesium-137 deposition and contamination of Japanese soils due to the Fukushima nuclear accident. Proc. Natl. Aced. Sci. USA (PNAS) **108**(49), 6 December 2011

Debo, T.N., Reese, A.J.: Municipal Stormwater Management, 2nd edn. CRC Press, Boca Raton, FL (2003)

Unoki, K (鵜木啓二)., Tada, D (多田大嗣)., Nakamura, K (中村和正)., Ikeda, H (池田晴彦)., Hosokawa, H (細川博明).: 畑地流域における USLE による土砂流出量の推定. 第 53 回 (平成 21 年度)北海道開発技術研究発表会, February 2010. (in Japanese)

Walton, A.: The distribution in soils of radioactivity from weapons tests. J. Geophys. Res. **68**(5), 1485–1496 (1963)

Wischmeier, W.H., Smith, D.D.: Predicting rainfall erosion losses – a guide to conservation planning. U.S. Department of Agriculture, Agriculture handbook No. 537 (1978)

Abukuma River Basin Risk Reduction Committee (阿武隈川圏域総合流域防災協議会): 阿武隈川圏域の水害・土砂災害対策における課題と当面の進め方、平成 18 年 9 月 5 日 (2006). (in Japanese)

Radioactive Waster Management Center: Environmental Parameters Series 1: Transfer Factors of Radionuclides from Soils to Agricultural Products (1988)

Japan Meteorological Agency: AMeDAS; Previous data. http://www.data.jma.go.jp/obd/stats/etrn/index.php

Mizugaki, S., Onda, Y., Fukuyama, T., Koga, S., Asai, H., Hisamatsu, S.: Estimation of suspended sediment sources using 137Cs and 210 Pb$_{ex}$ un unmanaged Japanese cypress plantation watersheds in sourthern Japan. Hydrol. Process. **22**, 4519–4531 (2008)

National Institute for Environmental Studies of Japan (NIES), G-CIEMS(Grid-Catchment Integrated Modeling System (多媒体環境モデリング) (2012). (in Japanese)

Charging and Discharging Characteristics of a Quantum Well with an Initially Lorentzian Wave Packet Incident on a DTM Type Potential

Youichirou Kubo and Norifumi Yamada[✉]

Information Science, Graduate School of Engineering,
University of Fukui, 3-9-1 Bunkyo, Fukui, Fukui 910-8507, Japan
yamada@u-fukui.ac.jp

Abstract. We study how non-interacting electrons accumulate in the quantum well region of a potential structure (simplified potential structure of Direct Tunneling Memory) after they start moving toward the potential as a quantum mechanical wave packet. The probability $P_L(t)$ of finding an electron in the well region obtained with an initially Lorentzian wave packet behaves differently in several respects from the probability $P_G(t)$ obtained with an initially Gaussian wave packet. Surprisingly, $P_L(t)$ can increase rather than decrease in the decay (discharging) process, implying that the electrons leaking from the well periodically change the direction of motion and move backward to the well. This counterintuitive "backflow effect" is caused by the quantum mechanical interference between the waves with two wave numbers k_0 and k_r, where k_0 is the most dominant wave number of the initial wave packet, and k_r is the resonance wave number of the well. We also discuss similarity and difference between $P_L(t)$ and $P_G(t)$ in the buildup (charging) process.

Keywords: Quantum dynamics · Non-exponential decay · Direct tunneling memory

1 Introduction

An important piece that governs the operating speed of quantum devices that use resonant tunneling phenomenon in quantum well structures is the time response characteristics of the quantum well. Direct tunneling memory (DTM) [1,2] is an example of such a quantum device. The memory's 0 and 1 states correspond respectively to the absence and the presence of electrons in its quantum well region. The writing time and the retention time of the memory are related, respectively, to the buildup characteristics and the decay characteristics of $P(t)$, which is the probability to find an electron in the quantum well region of DTM at time t. The behavior of $P(t)$ was studied when the electron incident on the DTM type potential is characterized as a Gaussian wave packet [3]. For related but different potential structures such as double barrier structures, many references exist which study $P(t)$ with Gaussian incident packets [4–6].

© Springer Science+Business Media Singapore 2016
S.Y. Ohn and S.D. Chi (Eds.): AsiaSim 2015, CCIS 603, pp. 118–128, 2016.
DOI: 10.1007/978-981-10-2158-9_11

An incident wave packet is not necessarily a Gaussian. Little is known about $P(t)$ when a non-Gaussian wave packet is incident on a DTM type potential. In this paper, we consider a wave packet of Lorentzian shape as an example of non-Gaussian incident packets, and study the resultant $P(t)$ in detail. We will show the following: The obtained $P(t)$ for a Lorentzian incident packet (hereafter, referred to as $P_L(t)$) shows a rapid increase and a slow decay as in the case of $P_G(t)$ for a Gaussian incident packet. However, the functional forms are quite different between $P_L(t)$ and $P_G(t)$. In the buildup (charging) process where both $P_G(t)$ and $P_L(t)$ show a rapid increase, $\log P_L(t)$ is a convex function, while $\log P_G(t)$ is a concave function. In the decay (discharging) process, $P_G(t)$ decreases exponentially, whereas $P_L(t)$ shows a complicated behavior where the exponential decay is modulated by an oscillation. Surprisingly, $P_L(t)$ increases periodically in the decay process as the result of the oscillation. This implies that the electrons leaking from the well periodically change the direction of motion and move backward to the well to re-charge it.

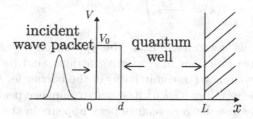

Fig. 1. A simplified potential profile of DTM and an incident wave packet.

Although an actual DTM is a three dimensional object and the electrons interact each other, we consider a simplified model where non-interacting electrons described as a single particle wave packet move in one-dimensional space and the wave packet is scattered by the simplified potential $V(x)$ of a DTM structure as shown in Fig. 1, where the value of $V(x)$ is 0, V_0, 0, and ∞ for $x < 0$, $0 < x < d$, $d < x < L$, and $x > L$, respectively. If the readers not familiar with quantum mechanics find it difficult to understand the electron's accumulation in the well in terms of $P(t)$, simply multiply a large number N with $P(t)$ to convert the probability to the number of electrons in the well at time t.

2   Probability $P(t)$ to Find an Electron in Quantum Well

We consider the case where a wave packet is incident on the simplified potential from the left as shown in Fig. 1. The probability to find an electron in the well region at time t is given by

$$P(t) = \int_d^L |\psi(x,t)|^2 dx, \tag{1}$$

where $\psi(x,t)$ is the electron's wave function, which is the solution to the time dependent Schrödinger equation

$$i\hbar\frac{\partial\psi(x,t)}{\partial t} = -\frac{\hbar^2}{2m}\frac{\partial^2\psi(x,t)}{\partial x^2} + V(x)\psi(x,t) \qquad (2)$$

with a given initial condition $\psi(x,0)$, where m is the effective mass of the electron and $\hbar$ is the reduced Planck constant (or Dirac's constant). The wave function in the well region can be expressed in the following form as the superposition of plane waves:

$$\psi(x,t) = \frac{1}{\sqrt{2\pi}} \int_{-\infty}^{\infty} C(k)f(k)\sin(k(x-L))e^{-i\frac{\hbar k^2}{2m}t}dk, \qquad (3)$$

where $f(k)$, the weighting function for superposition, is the Fourier transform of the initial wave function, i.e.,

$$f(k) = \frac{1}{\sqrt{2\pi}} \int_{-\infty}^{\infty} \psi(x,0)e^{-ikx}dx. \qquad (4)$$

The function $C(k)$ is determined from the appropriate boundary conditions imposed on the wave functions in respective regions, and its modulus $|C(k)|^2$ represents the transparency (transmittivity) of the barrier for the plane wave of wave number k. The graph of $|C(k)|^2$ has many resonance peaks. We consider a situation where only one sharp resonance peak appears in the range of k where $f(k)$ takes non-negligible values (this is physically the most relevant situation). In this case, only one resonance peak contributes to the k integral in (3). We can then approximate $C(k)$ in the vicinity of $k = k_r$ as follows:

$$C(k) \simeq \frac{i\Delta_r}{k - k_r + i\Delta_r}C(k_r), \qquad \Delta_r = \frac{1}{\theta'(k_r)}, \qquad (5)$$

where θ is the phase of $C(k)$ $(= |C(k)|e^{i\theta(k)})$, the prime represents a k differentiation, and Δ_r is the half width at half maximum of the resonance.

The steps of obtaining $P(t)$ is as follows: We give an initial wave function $\psi(x,0)$, and use (4) to obtain $f(k)$. We also solve the time independent Schrödinger equation to obtain the steady state solutions, and determine the coefficient $C(k)$. We choose a resonance wave number k_r at which $|C(k)|^2$ takes a maximal value. We calculate $\theta'(k_r)$ numerically by using

$$\theta'(k_r) = \text{Im}\left(\frac{C'(k_r)}{C(k_r)}\right). \qquad (6)$$

In order for the requirement stated above (5) to be filled, we must choose the width δx of the incident packet in such a way that the following relation holds.

$$L \ll \delta x \ll \frac{1}{\Delta_r} \qquad (7)$$

With thus determined $f(k)$ and $C(k)$, we use (3) to calculate $\psi(x,t)$, and then use (1) to obtain $P(t)$, where the k and the x integrations are calculated numerically.

Let us, for now, consider the initial wave packet given by

$$\psi_G(x,0) = \frac{1}{\sqrt[4]{2\pi(\delta x)^2}} \exp\left[-\frac{(x-x_0)^2}{4(\delta x)^2} + ik_0 x\right]. \tag{8}$$

This is called a Gaussian wave packet, because $|\psi_G(x,0)|^2$ is a normal distribution. Equations (4) with (8) gives

$$f_G(k) = \sqrt[4]{\frac{2(\delta x)^2}{\pi}} \exp\left[-\delta x^2 (k-k_0)^2 - i(k-k_0)x_0\right], \tag{9}$$

where x_0 and k_0 are, respectively, the peak position and the central wave number of the initial wave packet, and δx is the standard deviation of the normal distribution. Figure 2 shows a typical example of $P_G(t)$ which we obtained by following the steps described above. The parameters used are: $x_0 = -10.0\,\mu m$, $\delta x = 0.5\,\mu m$, $k_0 = k_r \simeq 4.9482915\,nm^{-1}$, $d = 1.0\,nm$, $V_0 = 3.3\,eV$, $L = 295.984\,nm$, and $m = 0.33\,m_0$ (m_0 is the bare electron mass). The use of these parameter values ensures that the requirement stated above (5) is met (see Fig. 3). The value of Δ_r is $0.029532489\,pm^{-1}$.

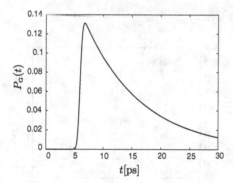
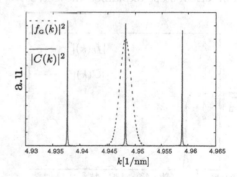

Fig. 2. The probability of finding an electron in the well region of the DTM type potential when the initial wave packet is given by (8).

Fig. 3. Only one sharp resonance peak is covered by the weighting function $f_G(k)$.

3 $P(t)$ in the Case of Lorentzian Incident Wave Packet

3.1 Lorentzian Incident Wave Packet and an Example of $P_L(t)$

We consider the initial wave packet given by

$$\psi_L(x,0) = \frac{1}{\sqrt{\pi}} \frac{\sqrt{\delta x}}{x - x_0 + i\delta x} e^{ik_0 x}, \tag{10}$$

where we call $\psi_L(x,0)$ given above a Lorentzian wave packet, because $|\psi_L(x,0)|^2$ is a Lorentz function $\delta x/\pi\{(x-x_0)^2+\delta x^2\}$. Substituting (10) into (4), we have

$$f_L(k) = \begin{cases} \sqrt{2\delta x}\exp\{-\delta x(k-k_0) - i(k-k_0)x_0\} & (k \geq k_0) \\ 0 & (k < k_0). \end{cases} \tag{11}$$

Not only for the Gaussian incident packet given by (8), but also for the Lorentzian incident packet given by (10), the wave number k_0 determines the velocity v_0 of the packet peak such that $v_0 = \hbar k_0/m$, and δx represents the width of the packet. It should be mentioned here that, although δx in (8) and that in (10) are both a measure of the "initial packet width," the two δx have slightly different meanings; δx in (8) is the standard deviation of the normal distribution $|\psi_G(x,0)|^2$, while δx in (10) is the half width at half maximum of the Lorentz function $|\psi_L(x,0)|^2$.

To make the comparison between $P_L(t)$ and $P_G(t)$ meaningful, we use the same parameter values as were used in Sect. 2. The choice of the value of k_0 is, however, exceptional. If we use the same value of k_0 as was used in Sect. 2, only a half of the resonance peak is covered by the weighting function $f_L(k)$. To avoid this, we use a slightly different value of k_0 (4.9475501 nm^{-1}) to make sure that the resonance peak is fully covered by $f_L(k)$ as shown in Fig. 4. With this careful choice of the parameter values, we obtain $P_L(t)$ shown in Fig. 5 by following the numerical steps explained in Sect. 2.

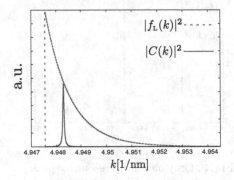

Fig. 4. Only one sharp resonance peak is covered by the weighting function $f_L(k)$.

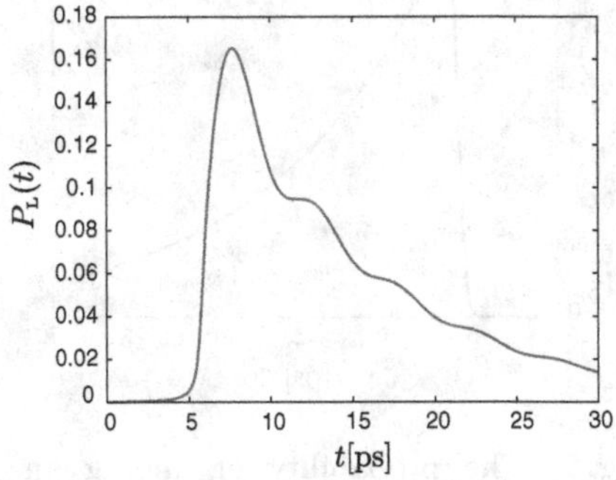

Fig. 5. The probability of finding an electron in the well region of the DTM type potential when the initial wave packet is given by (10).

3.2   Comparison Between $P_L(t)$ and $P_G(t)$

In Fig. 6, we compare $P_L(t)$ (solid line) and $P_G(t)$ (dashed line). The two curves share the same characteristics, i.e., they both increase rapidly and decay slowly. In details, however, there are several differences between them as pointed out below.

Figure 7 shows $\log P_L(t)$ and $\log P_G(t)$. It is clear from the figure that $P_G(t)$ shows an exponential decrease in the decay process. We have confirmed that the exponent of the exponential decay agrees with theory; the slope of $\log P_G(t)$ obtained from the numerical data, $-0.1025284\,\mathrm{ps}^{-1}$, is very close to the theoretical value $-0.1025318\,\mathrm{ps}^{-1}$ $(= -2\hbar k_0 \Delta_r/m)$. From Figs. 6 and 7, we may say that $P_L(t)$ shows a modulated exponential decay, where an exponential decay with the same negative exponent as in the case of $P_G(t)$ is modulated by an oscillation of a small amplitude.

Figure 8 is an enlarged view of the modulated exponential decay of $P_L(t)$. Surprisingly, $P_L(t)$ increases between $t = 11.56\,\mathrm{ps}$ and $t = 11.85\,\mathrm{ps}$. This implies that the electron moves to the right in this time interval. The same thing happens at later times as shown in Fig. 9. The temporary increase of the probability in the well region means that the electron leaking from the well temporarily moves backward to the well to re-charge it. This view is supplemented by Fig. 11, where the flux (probability current density) is shown. The electron is moving to the right when the flux takes a positive value, while it is moving to the left when it takes a negative value. Figure 11 shows that, in the time interval where the temporary increase is observed, the flux is positive, and thus the leaking electron is moving backward to the well. Although the temporary increase of the probability observed here is small, it can become much larger depending on the system parameter values as we will see in the next subsection. Note that a large temporary increase is of practical significance, because it would cause a memory error of DTM. The cause of the temporary increase is explored in Sect. 3.4.

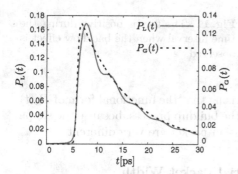

Fig. 6. The comparison between $P_L(t)$ in Fig. 5 and $P_G(t)$ in Fig. 2.

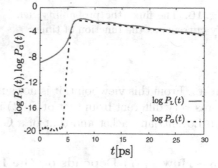

Fig. 7. Natural log plot of $P_L(t)$ and $P_G(t)$.

Let us return to Fig. 7 to see the buildup process where the two curves show rapid increase. We see there that $\log P_L(t)$ is a convex function, while $\log P_G(t)$ is a concave function. This implies that the functional forms in the buildup process are quite different between $P_G(t)$ and $P_L(t)$, although both curves in the buildup process do not look so different in Fig. 6. We consider that the functional form of $P(t)$ in the buildup process is sensitive to the functional form of the incident

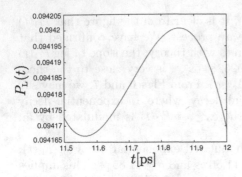

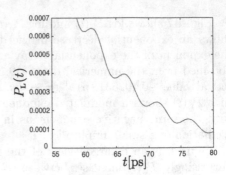

Fig. 8. In the decay process, a temporary increase of $P_L(t)$ is observed between $t = 11.56$ ps and $t = 11.85$ ps.

Fig. 9. Temporary increase of $P_L(t)$ at later times.

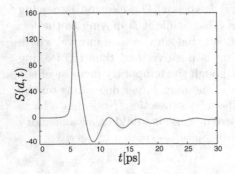

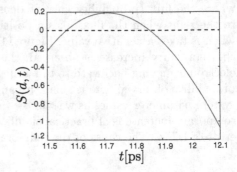

Fig. 10. The flux (the probability density current) as the function of time.

Fig. 11. The flux is positive during the time interval when the backflow effect is observed.

packet. From this viewpoint, it is not surprising that the functional form of $P_L(t)$ seems very different from that of $P_G(t)$ in the buildup process, because the shape of a Lorentzian packet and that of a Gaussian packet are very different.

3.3 How $P_L(t)$ Depends on the Initial Packet Width

The probability $P(t)$ of finding an electron in the quantum well region at time t depends on many system parameters. Here, among them, let us change the "width" δx of the initial wave packet to see how $P(t)$ is altered by the packet size, where δx represents the half width at half maximum of the initial probability density for a Lorentzian packet, while it represents the standard deviation of the initial probability density for a Gaussian packet. Figures 12, 13, and 14 show $P_L(t)$ for the case of $\delta x = 500$ nm, 1500 nm, and 3000 nm, respectively, where other parameter values are the same as those used in Fig. 5 except for x_0, which is $-100\,\mu$m in the present examples.

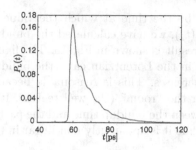

Fig. 12. $P_L(t)$ with $\delta x = 500\,\text{nm}$.

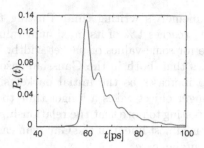

Fig. 13. $P_L(t)$ with $\delta x = 1500\,\text{nm}$.

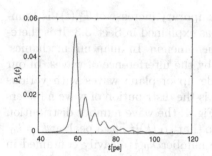

Fig. 14. $P_L(t)$ with $\delta x = 3000\,\text{nm}$.

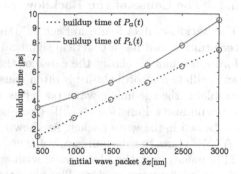

Fig. 15. The dependence of the buildup time on the initial packet width.

We first find that the temporary increase of $P_L(t)$ becomes more distinct as the packet width becomes large. This means that the experimental detection of the backflow effect would be easier if energetically monochromatic electrons are used as incident electrons (due to the position-momentum uncertainty principle in quantum mechanics, a large packet width means a small uncertainty in momentum, and thus a small uncertainty in energy of the electrons). As mentioned earlier, a large temporary increase of the probability is harmful to DTM. The numerical examples shown here warn that non-Gaussian incident packets could cause a fatal memory error; one would not notice this in numerical simulations if only Gaussian initial packets are used in the simulations.

The Figs. 12, 13, and 14 also show that the peak value of $P_L(t)$ becomes small as the packet width becomes large, which means that less electrons enter the well region as the incident electrons become monochromatic in energy (thus the experimental detection of the backflow effect becomes difficult if the packet size is too large). We also notice that $P_L(t)$ stands up sharply in the build-up process when the initial packet width is small. To characterize how sharply it stands up (i.e., how fast the well is filled with electrons), let us define the buildup time (charging time) by $t_2 - t_1$, where t_2 is the time at which $P_L(t)$ reaches its maximum value and t_1 is the time at which it starts to take an appreciable value. The buildup time has a practical meaning for DTM, because it corresponds to the memory's writing time. Putting t_1 to be the time at which the value of $P_L(t)$ reaches 1% of its maximum value $P_L(t_2)$, we have calculated the buildup time for some values of packet width. The result is shown in Fig. 15. The figure shows that, both in the Gaussian case and in the Lorentzian case, the buildup time increases as the initial packet size increases. This is reasonable because a longer object takes a longer time to enter the "room" (the well region). It is interesting to note that the relationship between the buildup time and the packet size is almost linear in the Gaussian case, while it is apparently non-linear in the Lorentzian case.

3.4   The Cause of the Backflow Effect

The backflow effect found in Sect. 3.2 (the temporary increase of $P_L(t)$) is counterintuitive, and has a practical significance as explained in Sect. 3.3. It is therefore important to clarify the cause of the phenomenon. In quantum mechanics, an oscillation of probability is often caused by the interference of waves. In our problem, the incoming wave packet is made up of plane waves with various wave numbers. Function $|f_L(k)|^2$ (see Fig. 4) is the distribution of wave numbers contained in the wave packet. As shown in Fig. 4, the wave number distribution $|f_L(k)|^2$ starts to take non-zero value at k_0, so that $|f_L(k)|^2$ has a peak at $k = k_0$. This means that the plane wave with wave number k_0 is heavily contained in the incoming wave packet. The potential structure, on the other hand, allows the plane wave with wave number k_r to be preferably transmitted into the well, because $|C(k)|^2$ has a resonance peak at $k = k_r$. As the result, the overall weighting factor $C(k) f(k)$ on the right-hand side of (3) is doubly peaked at $k = k_0$ and k_r. Therefore, the wave in the well region has two dominant plane wave components, that is, those with wave number k_0 and k_r. This view leads us to

$$|ae^{ik_0x-i\omega_0t} + be^{ik_rx-i\omega_rt}|^2 \tag{12}$$

as a rough expression for the probability density in the well, where $\omega_0 = \hbar k_0^2/(2m)$ and $\omega_r = \hbar k_r^2/(2m)$, and a and b are constants. Equation (12) gives an interference term $e^{i(\omega_r-\omega_0)t}$. Therefore, (12) and hence $|\psi_L(x,t)|^2$ would oscillate in time with a period T_{os} given by

$$T_{os} = \frac{2\pi}{\omega_r - \omega_0}. \tag{13}$$

When $|\psi_L(x,t)|^2$ oscillates in time with period T_{os}, its integration over position x, namely, $P_L(t)$ would also oscillate in time with period T_{os}.

We have numerically confirmed that the above view is correct. Figures 16, 17, 18, and 19 show the oscillating part of $P_L(t)$, which we obtained by changing the system parameter values. Equation (13) gives 4.88 ps, 3.45 ps, 2.88 ps, and 0.74 ps as the oscillation periods expected for Figs. 16, 17, 18, and 19, respectively. It is clear that the actual oscillation periods obtained from the numerical data agree very well with the theoretical values given by (13). We may thus conclude that the oscillations observed in $P_L(t)$ is caused by the quantum mechanical

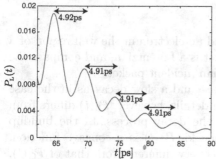

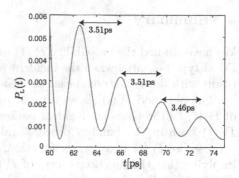

Fig. 16. Oscillatory behavior $P_{\mathrm{L}}(t)$ in the decay process.

Fig. 17. Oscillatory behavior $P_{\mathrm{L}}(t)$ in the decay process.

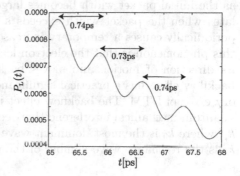

Fig. 18. Oscillatory behavior $P_{\mathrm{L}}(t)$ in the decay process.

Fig. 19. Oscillatory behavior $P_{\mathrm{L}}(t)$ in the decay process.

interference between the waves with wave numbers k_0 and k_{r}. The cause of the counterintuitive backflow effect is, therefore, a purely quantum mechanical one.

The parameter values we used are as follows: As to Figs. 16 and 17: $\delta x = 3000\,\mathrm{nm}$ and $k_{\mathrm{r}} = 4.94829150000208\,\mathrm{nm}^{-1}$ for both figures; $k_0 = 4.9475501\,\mathrm{nm}^{-1}$ for Fig. 16 and $k_0 = 4.94724240001316\,\mathrm{nm}^{-1}$ for Fig. 17. As to Figs. 18 and 19: $\delta x = 500\,\mathrm{nm}$ for both figures; $k_{\mathrm{r}} = 4.714319700047028\,\mathrm{nm}^{-1}$ and $k_0 = 4.7130\,\mathrm{nm}^{-1}$ for Fig. 18; $k_{\mathrm{r}} = 5.160935000005426\,\mathrm{nm}^{-1}$ and $k_0 = 5.156214700132405\,\mathrm{nm}^{-1}$ for Fig. 19. For Figs. 16, 17, 18, and 19, $m = 0.33\,m_0$, $x_0 = -100000\,\mathrm{nm}$, $V_0 = 3.3\,\mathrm{eV}$, $d = 1\,\mathrm{nm}$, $L = 295.984\,\mathrm{nm}$.

Incidentally, although not in the context of DTM, the quantum backflow effect has also been known for superposition of Gaussian wave functions (see, for example, Ref. [7]), where the size of the effect is relatively small. The present study shows that the backflow effect can be made more distinct with a Lorentzian initial wave function. It also shows that the quantum backflow effect, which is often studied in a purely academic context, can also be a practical issue in device applications.

4 Summary

We have studied the probability $P_L(t)$ to find an electron in the well region of a DTM type potential when the incident packet is a Lorentzian, and compare the result with the well-studied case of a Gaussian incident packet.

The obtained $P_L(t)$ shows a rapid increase and a slow decay as in the case of $P_G(t)$ for the Gaussian incident packet. In details, however, $P_L(t)$ differs from $P_G(t)$ both in the buildup process and in the decay process. In the buildup process, $\log P_L(t)$ is a convex function, while $\log P_G(t)$ is a concave function, implying that the functional form of $P_L(t)$ is very different from that of $P_G(t)$. The buildup time shows a linear dependence on the packet width in the Gaussian case, but a non-linear dependence in the Lorentzian case. In the decay process, $P_L(t)$ shows a modulated exponential decay. The modulation amplitude increases as the initial packet width becomes large, and $P_L(t)$ eventually begins to oscillates when the packet width exceeds a certain value. The oscillation of $P_L(t)$ periodically causes a temporary increase of the probability. A physical view of this phenomenon is that the electron leaking from the well periodically changes its direction of motion and moves backward to the well to re-charge it. The backflow effect has a practical significance, because it could cause a fatal memory error of DTM. The backflow effect is counterintuitive and is caused by the quantum mechanical interference between the waves with wave numbers k_0 and k_r, where k_0 is the most dominant wave number of the initial wave packet, and k_r is the resonance wave number of the well.

References

1. Kuo, C., King, T.J., Hu, C.: Direct tunneling RAM (DT-RAM) for high density memory applications. IEEE Electron Dev. Lett. **24**, 475–477 (2003)
2. Tsunoda, K., Sato, A., Tashiro, H., Nakanishi, T., Tanaka, H.: Improvement in retention/program time ratio of direct tunneling memory (DTM) for low power SoC applications. IEICE Trans. Electron. **88-C**(4), 608–613 (2005)
3. Kubo, Y., Yamada, N.: Novel method for simulating quantum buildup process in the potential well of a DTM potential. In: Proceedings of 33rd JSST Annual Conference (JSST 2014), pp. 70–71 (2014)
4. Inaba, H., Nakagawa, J., Kurosawa, K., Okuda, M.: Dynamics of resonant tunneling in double-barrier structures with trapezoidal potential profile. Jpn. J. Appl. Phys. **30**, L544–L546 (1991)
5. Hauge, E.H., Støvneng, J.A.: Time-dependent resonant tunneling of wave packets in the tight-binding model. Phys. Rev. B **44**, 13582–13594 (1991)
6. Gong, J., Liang, X.X., Ban, S.L.: Tunneling time of electronic wave packet through a parabolic quantum well with double barrier. Phys. Stat. Sol. (B) **244**, 2064–2071 (2007)
7. Yearsley, J.M., Halliwell, J.J.: An introduction to the quantum backflow effect. J. Phys.: Conf. Ser. **442**, 012055 (2013)

Author Index

Printed in the United States
By Bookmasters